Introduction
to Polymer
Viscoelasticity

Introduction

to Polymer

Viscoelasticity

JOHN J. AKLONIS
University of Southern California

WILLIAM J. MacKNIGHT
University of Massachusetts

MITCHEL SHEN
University of California, Berkeley

WILEY-INTERSCIENCE
a division of John Wiley & Sons, Inc.
New York   London   Sydney   Toronto

Library of Congress Cataloging in Publication Data:

Aklonis, John J.
Introduction to polymer viscoelasticity.

Includes bibliographies.
1. Polymers and polymerization. 2. Viscoelasticity.
I. MacKnight, William J., joint author. II. Shen,
Mitchel C., 1938–   joint author. III. Title.

TA455.P58A35   620.1'92'32   72-473
ISBN 0-471-01860-0

Printed in the United States of America

10 9 8 7 6 5 4 3 2 1

To A. V. Tobolsky
who first introduced us
to the mysteries of polymer viscoelasticity
and to the meaning of scientific research.

Preface
◆◆◆◆◆◆◆◆◆◆

The viscoelastic response of polymeric materials is a subject which has undergone extensive development over the past twenty years and still accounts for a major portion of the research effort expended. It is not difficult to understand the reason for this emphasis in view of the vast quantities of polymeric substances which find applications as engineering plastics and the still greater volume which are utilized as elastomers. The central importance of the time and temperature dependence of the mechanical properties of polymers lies in the large magnitudes of these dependencies when compared to other structural materials such as metals. Thus an understanding of viscoelastic behavior is fundamental for the proper utilization of polymers.

Viscoelasticity is a subject of great complexity fraught with conceptual difficulties. It is possible to distinguish two basic approaches to the subject which we shall designate as the continuum mechanical approach and the molecular approach. The former attempts to describe the viscoelastic behavior of a body by means of a mathematical schema which is not concerned with the molecular structure of the body, while the latter attempts to deduce bulk viscoelastic properties from molecular architecture. The continuum mechanical approach has proven to be very successful in treating a large number of problems and is of very great importance. However, it is not our intention to treat this approach rigorously in this text. Rather, we shall be concerned with the molecular approach and attempt to present a basic foundation upon which the reader can build. The fundamental difficulty encountered with the molecular approach lies in the fact that polymeric materials are large molecules of very complex structures. These structures are too complex, even if they

vii

were known in sufficient detail which, in general, they are not, to lend themselves to mathematical analysis. It is therefore necessary to resort to simplified structural models, and the results deduced from these are limited by the validity of the models adopted.

Several excellent treatments of molecular viscoelasticity are available. (See the references of Chapter 1.) The book by Professor Ferry, in particular, is an exhaustive and complete exposition. The question may then be asked, why the necessity for still another text and one restricted to bulk amorphous polymers, at that? Such a question must send each of the authors scurrying in quest of an "apologia pro vita sua." The answer to the question lies in the use of the word "introduction" in the title. What we have attempted to do is to provide a detailed grounding in the fundamental concepts. This means, for example, that all derivations have been presented in great detail, that concepts and models have been presented with particular attention to assumptions, simplifications, and limitations, and that problems have been provided at the end of each chapter to illustrate points in the text. The level of mathematical difficulty is such that the average baccalaureate chemist should be able to readily grasp it. Where more advanced mathematical techniques are required, such as transform techniques, the necessary methods are developed in the text.

Having attempted to delineate what this book is, it may be well to remind the reader what it is not. First of all, it is not a complete treatment—lacking among other topics discussions of crystalline polymers, solution behavior, melt rheology, and ultimate properties. It is also not written from the continuum mechanics approach and thus is not mathematically sophisticated. Finally, it is not a primer of polymer science. Familiarity with the basic concepts of the field is presumed.

The authors' first acquaintance with the literature of viscoelastic behavior in polymers evoked a response much like that experienced by neophytes in the literary arts on a first reading of "Finnegan's Wake," by James Joyce. It is immediately apparent that one is in the presence of a great work, but somehow it will be necessary to master the language before appreciation, let alone understanding, may be achieved. Recognizing the nature of the problem, Joycean scholars came to the rescue with works of analysis to provide a skeleton key to "Finnegan's Wake." Proper utilization of this skeleton key will open the door to an understanding of that forbidding masterwork. It was thus our intent to provide a similar skeleton key to the literature of molecular viscoelasticity. How well we have succeeded must be left to the judgment of our readers.

We are grateful to the students at our respective institutions who have suffered our attempts to present the material in this text in coherent form at various stages of development. Their criticisms and suggestions have led to significant improvements. We are also grateful to Mrs. William Jackson, who translated many rough sketches into finished drawings. It is hardly to be expected that a work of this nature could be free from errors. We have attempted to eliminate as many as possible but, of course, bear full responsibility for those remaining.

Los Angeles, California   John J. Aklonis
William J. MacKnight
Mitchel Shen

Contents
◆◆◆◆◆◆◆◆◆◆◆

xi

Introduction to Polymer Viscoelasticity

1 Introduction
◆◆◆◆◆◆◆◆◆◆◆◆◆◆◆◆◆◆◆◆◆

The subject matter of this book consists of the response which certain classes of substances make when they are subjected to external forces of various kinds. The classes of substances to be treated are those known as viscoelastic bodies. As the name implies, these materials respond to external forces in a manner intermediate between the behavior of an elastic solid and a viscous liquid. In order to set the stage for what follows it is necessary to quantitatively define the types of forces to which the viscoelastic bodies are to be subjected.

Consider first the motion of a rigid body in space. This motion can be thought of as consisting of translational and rotational components. If no forces act on the body, it will maintain its original state of motion indefinitely in accordance with Newton's first law of motion. However, if a single force or a set of forces whose vector sum is nonzero act on the body, it will experience acceleration or a change in its state of motion. Consider, however, the case where the vector sum of forces acting on the body is zero and the body experiences no change in either its translational or rotational component of motion. In such a condition, the body is said to be *stressed*. If the requirement of rigidity is removed, the body will in general undergo a deformation as a result of the application of these balanced forces. If this occurs, the body is said to be *strained*. It is the relationship between stress and strain which is our main concern. Depending on the types of stress and

Figure 1-1. A Simple Extension Test Used to Define the Modulus.

strain applied to a body, it is possible to use these quantities to define new quantities which relate to the mechanical strength of the body. These new quantities are called moduli. In order to understand the physical meaning of the modulus of a solid, consider the following simple experiment.

Suppose we have a piece of rubber (e.g., polyisobutylene) $\frac{1}{2}$ cm × $\frac{1}{2}$ cm × 4 cm and a piece of plastic (e.g., polystyrene) of the same dimensions. The experiment to be performed consists of suspending a weight (applying a force) of, say 1 kg, from each material as shown in Figure 1-1. As is obvious, the deformation of the rubber will be much greater than that of the plastic. Using this experiment we define a modulus as a quantity proportional to the applied force divided by the change in length, i.e.,

$$M \propto \frac{\text{Force}}{\Delta L} \tag{1-1}$$

Since ΔL is much larger for the rubber than for the plastic, from equation (1-1) it is clear that the modulus of the rubber is much lower than the modulus of the plastic. Thus the particular modulus defined in equation (1-1) defines the resistance of a material to elongation.

Further experimentation, however, reveals that the situation is more complicated than is initially apparent. If, for example, one were to carry out the test on the rubber at liquid nitrogen temperature, one would find that this "rubber" would undergo a much smaller elongation than with the same force at room temperature. In fact, the extension would be so small as to be comparable to the extension exhibited by the plastic at room temperature. A more dramatic demonstration of this effect is obtained by immersing a rubber ball in liquid nitrogen for several minutes. The cold ball, when bounced, will no longer have the characteristic properties of a rubbery object but, instead, will be indistinguishable from a hard sphere made of plastic.

On the other hand, if the piece of plastic is heated in an oven to 130°C and then subjected to the modulus measurement, it is found that a much larger elongation, comparable to the elongation of the rubber at room temperature, results.

These simple experiments indicate that the modulus of a polymeric material is not invariant, but is a function of temperature, i.e., $M = M(T)$.

An investigation of the temperature dependence of the modulus of our two samples is now possible. At temperature T_1 we will measure the modulus as before, then increase the temperature to T_2, etc. Schematic data from such an experiment are plotted in Figure 1-2. The temperature dependence of the modulus is so great that it must be plotted on a logarithmic scale. (This great variation in modulus presents experimental problems which will be treated subsequently.) The modulus at the lowest temperature measurement (plastic) has been arbitrarily defined as 1.0 in the figure and other moduli are given relative to this value. The area between the vertical dashed lines represents room temperature and, consistent with the first experiment, we find that the plastic has a high modulus while the rubber has a relatively low modulus. Upon cooling, the modulus of the rubber rises markedly by about four orders of magnitude indicating that the rubber, at lower temperatures, behaves like a

Figure 1-2. Schematic Modulus-Temperature Curves for a Rubber and a Plastic over a Broad Temperature Range.

plastic. Another drastic change in modulus is evidenced at higher temperatures for the rubber; here the material is becoming softer, as indicated by further decreases in modulus. This behavior will be discussed in detail in Chapter 3. The modulus-temperature behavior for the plastic is seen to be quite similar to that of the rubber except that the changes occur at higher temperatures, resulting in the high modulus observed at room temperature. At 135°C it is clear that the modulus of this material is that of a rubber, agreeing with the results of one of the earlier "experiments" in this discussion.

One more type of deformational experiment remains to be discussed. Consider a material like pitch or tar, which is used as a roof coating and is applied at elevated temperatures. Our test will be similar to the standard experiment done above utilizing the same size sample at room temperature. First we suspend the 1 kg weight from the sample and observe the small resultant extension. According to equation (1-1), the modulus calculated is high. However, if the weight is removed and the sample left suspended in this vertical position for several days, it is observed that the rather weak force exerted by gravity results in a considerable elongation of the sample. Now application of equation (1-1) gives a very low value for the modulus. Thus the modulus measurement on the short time scale of minutes resulted in a high value while the modulus measurement on the longer time scale of days resulted in a low modulus. This apparent discrepancy is accounted for by realizing that the modulus is a function of time as well as temperature; this will be found to be the case generally when considering polymeric systems. Strictly then, the measurements spoken of earlier in this chapter and depicted schematically in Figure 1-2 should have some time associated with each modulus value. (Time represents the duration between the application of the force and the measurement of the extension fraction.) It is convenient to pick the same constant time for all measurements so one might consider the constant time factor in Figure 1-2 to be 10 seconds.

As is evident from the above discussion, it should be possible to measure the behavior of a material as a function of time at constant temperature. A schematic modulus-time behavior is shown in Figure 1-3. Here again the maximum value of the modulus is normalized

Figure 1-3. Schematic Modulus-Time Curve for a Polymer at Constant Temperature.

to 1.0 (log 1.0 = 0). The modulus is seen to fall from its initial high value by about three orders of magnitude to a modulus indicative of a rubber and, after evidencing a plateau, fall again. The ordinate here is log t; at this temperature an experiment lasting for 1 to 30 minutes would characterize this material as a plastic. However, to an experiment lasting 10^8 minutes (200 years), the material would "look like" a rubber. Longer measurements would correspond to still softer materials. Methods for obtaining curves of the type of Figure 1-3 will be discussed in Chapter 3 as well as methods of converting from modulus-time to modulus-temperature behavior and vice versa.

Another experiment is often carried out in laboratories dealing with the physical properties of polymers. This is the determination of the temperature at which the material properties change from those of a plastic to those of a rubber. This temperature is known as the glass transition temperature and is a characteristic property of each substance. In Figure 1-2, for example, it is clear that at about 100°C, the modulus of the plastic exhibits a steep decrease. Careful analysis of the curve in this region, however, indicates no abrupt change in modulus but rather a smoothly varying change. From this experiment, it would seem that the glass transition occurs over a range of temperatures rather than at a single temperature. Experimentally, it has been found that the coefficient of expansion of a substance undergoes an abrupt change in the region of the glass transition. The temperature at this abrupt change, in fact is defined as the glass transition temperature T_g.

Figure 1-4. Schematic of Data Used to Determine the Glass Transition Temperature of a Polymer.

An example of the data obtained in the determination of a glass transition temperature is presented in Figure 1-4. The volume of a sample is measured as a function of temperature, care being taken to change the temperature slowly. Experimentally one often employs a technique which is based on Archimedes' Principle. The temperature where there is a change in slope (due to a discontinuity in the coefficient of expansion) is taken as T_g.

References

1. T. Alfrey, *Mechanical Behavior of High Polymers*, Interscience, New York, 1948.

2. C. Zener, *Elasticity and Anelasticity of Metals*, University of Chicago Press, Chicago, 1948.

3. S. Timoshenko and J. N. Goodier, *Theory of Elasticity*, McGraw-Hill, New York, 1951.

4. E. N. da C. Andrade, *Viscosity and Plasticity*, Chemical Publishing Co., New York, 1951.

5. B. Gross, *Mathematical Structure of the Theories of Viscoelasticity*, Hermann & Cie, Paris, 1953.

6. F. R. Eirich, Ed., *Rheology, Theory and Applications*, Vol. 1, Academic Press, New York, 1956.

7. A. V. Tobolsky, *Properties and Structure of Polymers*, Wiley, New York, 1960.

8. J. D. Ferry, *Viscoelastic Properties of Polymers*, Wiley, New York, 1970.

9. L. E. Nielsen, *Mechanical Properties of Polymers*, Rheinhold, New York, 1962.

10. F. Bueche, *Physical Properties of Polymers*, Interscience, New York, 1962.

11. O. H. Varga, *Stress-Strain Behavior of Elastic Materials*, Interscience, New York, 1966.

12. A. V. Tobolsky and H. F. Mark, Eds., *Polymer Science and Materials*, Wiley, New York, 1971.

13. I. M. Ward, *Mechanical Properties of Solid Polymers*, Wiley, New York, 1971.

2 Phenomenological Treatment of Viscoelasticity

In order to put the concepts discussed in the Introduction on a more quantitative basis and to better understand some of the added complications which arise from the time dependence of the modulus, several methods commonly used to make physical measurements on polymeric systems will be discussed. First, however, it is necessary to rigorously define the parameters to be derived from the experimental data.

A. Elastic Moduli

Perhaps the simplest deformation which can be applied to a sample is uniaxial tension or compression, pictured in Figure 2-1a. This is the type of deformation mentioned in the Introduction. However, our previous concept of force is modified slightly in this more quantitative discussion to become the stress*

---

* In this context, both stress and strain are defined considering very small deformations. Under these conditions, for example, the cross sectional area of the sample before the deformation is virtually equal to that after. For large deformations, "true" and "engineering" functions are defined in texts such as references 1 and 2.

Figure 2-1. Tension and Shear in a Three-Dimensional Body.

defined as

$$\sigma_t = \frac{F}{AB} \tag{2-1}$$

where the subscript t represents uniaxial tension and the other symbols are defined in the figure. σ_t has the units of dynes per cm², newtons per meter², or PSI. It should be noticed that static definitions are being considered here. (The parameters do not vary with time.) These initial considerations will shortly be generalized to time-dependent functions.

The application of a stress to a real body will result in a strain as discussed in the Introduction. The fractional extension is defined as the tensile strain. Thus ε, the tensile strain resulting from the application of a uniaxial stress, is given as

$$\varepsilon = \frac{\Delta C}{C} \tag{2-2}$$

and it is clear that ε is dimensionless. In terms of classical physics, the tensile modulus E and the tensile compliance D are defined by equation (2-3).

$$E = \frac{\sigma_t}{\varepsilon} = \frac{1}{D} \qquad (2\text{-}3)$$

Again it must be stressed that this discussion is strictly limited to time-independent phenomena.

The second type of deformation is illustrated in Figure 2-1b and is called simple shear. The application of the force F will result in the deformation shown by the dashed lines. Here the stress σ_s is

$$\sigma_s = \frac{F}{AB} \qquad (2\text{-}4)$$

where the subscript s denotes shear. The shear strain is given as

$$\gamma = \frac{\Delta X}{C} = \tan \theta \qquad (2\text{-}5)$$

where θ is the angle shown and is related to $\Delta X/C$ by simple trigonometry. The shear modulus G and the shear compliance J are defined using the relationship

$$G = \frac{\sigma_s}{\gamma} = \frac{1}{J} \qquad (2\text{-}6)$$

From the theory of elasticity of isotropic solids, one has the relationship between E and G and D and J that

$$E = 2(1 + \mu)G \quad \text{or} \quad J = 2(1 + \mu)D \qquad (2\text{-}7)$$

where μ is Poisson's ratio defined as

$$\mu \equiv \frac{1}{2}\left[1 - \left(\frac{1}{V}\right)\frac{dV}{d\varepsilon} \right] \qquad (2\text{-}8)$$

Here, the fundamental relationship between shear and tensile properties is demonstrated. Equation (2-7) is derived in Appendix 1 of this chapter. For the special case of an incompressible material, $(\partial V/\partial \varepsilon) = 0$, $\mu = 0.5$, and one obtains the results

$$E = 3G \quad \text{or} \quad 3D = J$$

Experimentally μ is quite close to 0.5 for a rubber but is lower, in the range of 0.2 to 0.3, for some plastics and still lower for certain heterogeneous materials.

B. Transient Experiments

Consideration of the time dependence of relaxation phenomena adds additional complications. The value of a measured modulus or compliance will very definitely depend on the exact manner in which the experiment is carried out. As an example, consider the following experiments. First, a polymer is subjected to a constant uniaxial stress σ_1 for one hour; this perturbation results in some measurable strain, say ε (1 hour). In a second experiment, however, an identical sample is subjected to sufficient stress to result in the same strain ε (1 hour) immediately upon application of the stress. Then the stress is decreased so that the strain remains constant at ε (1 hour). The value of the stress after 1 hour in the second experiment is defined as σ_2. In general σ_1 and σ_2 will not be the same, the stress associated with the constant strain experiment being lower. However, since the strains are the same, the two "modulus" values calculated from equation (2-3) are different. Consequently one must explicitly state the method in which a parameter is to be measured in its definition. Fortunately, physics allows us to relate most parameters obtained from different experiments so that a knowledge of some of them will suffice to define all the others.[3-7] Some of these transformations will be treated later in this chapter.

In Chapter 1, we dealt with hanging a weight on a sample and observing the extension. This is a crude form of a creep experiment. In such an experiment the sample is subjected to constant stress and its strain is measured as a function of time. The shear creep compliance $J(t)$ is defined as

$$J(t) = \frac{\gamma(t)}{\sigma_{s_0}} \qquad (2-9)$$

where σ_{s_0} is the constant shear stress and the tensile creep compliance

$$D(t) = \frac{\varepsilon(t)}{\sigma_{t_0}} \qquad (2-10)$$

where σ_{t_0} is the constant tensile stress.

Figure 2-2. A Simple Apparatus to Measure Shear Creep.

Experimentally one might fasten a cubic sample to two light aluminum plates as shown in Figure 2-2. The sample would then be placed into a constant temperature environment where it would be allowed to come to thermal equilibrium before the start of the experiment (provision having been made to eliminate the stress due to the unloaded weight pan and string). When the experimenter is satisfied that the sample is at equilibrium, he places the appropriate weights on the weight pan and observes the deformation of the sample by noting the position of the weight pan with respect to the scale behind the apparatus. It should be clear that the shear strain $\gamma(t)$ as a function of time is easily calculated from these data and a knowledge of the original sample dimensions. The force exerted by the weights is used to calculate the constant shear stress and then the shear creep compliance $J(t)$ is calculated using the expression (2-9). It must be emphasized that this experimental setup is among the crudest imaginable and will give correspondingly imprecise results unless considerable refinement is made. The apparatus of Plazek[8] represents what is presently among the most sophisticated instrumentation used for measuring creep compliances.

Another common type of experiment, called stress relaxation, may be considered. Here as in experiment two above, the strain is maintained constant and the stress is measured as a function of time. A simple stress relaxation balance is shown in Figure 2-3. The sample, in the form of a strip, is first attached to the upper clamp and then the forces on either side of the fulcrum balanced by adjusting the counter weight (B). Next the lower clamp is attached and the sample raised to the temperature desired and thermal equilibrium established. (Usually the sample is raised to a temperature about 10 degrees above the test temperature or to a temperature above the glass temperature to allow possible "locked in" strains to relax before approach to the test temperature is made). At the start of the experiment, the strain is introduced quickly via the micrometer with the balance beam jammed against the upper stop. The application of weight to the balance pan will eventually cause the beam to move down from the upper stop toward the lower stop. This is the equilibrium point when the forces on either side of the fulcrum are balanced and simple calculations yield the stress at nearly constant strain. (The strain varies *minutely* during the measurement). This experiment measures $E(t)$, the tensile relaxation modulus,

$$E(t) = \frac{\sigma_t(t)}{\varepsilon_0} \qquad (2\text{-}11)$$

with ε_0 being the constant tensile strain.

A	SAMPLE
B	COUNTER WEIGHT
C	FULCRUM
D	WEIGHT PAN
E	STOPS FOR BEAM
F	MICROMETER
G	BEAM
H	CLAMPS

Figure 2-3. A Simple Apparatus to Measure Tensile Stress Relaxation.

Similarly, a shear stress relaxation experiment would measure $G(t)$, the shear stress relaxation modulus.

$$G(t) = \frac{\sigma_s(t)}{\gamma_0} \qquad (2\text{-}12)$$

with γ_0 being the constant shear strain.

This experiment has also been depicted in a rather simplified form. Usually the mechanical mechanism is supplanted by electronics which, in essence, perform the same tasks.

It must be noted that equations of the form of (2-3) and (2-6) relating moduli and compliances are no longer applicable to our new time-dependent functions, since

$$E(t) = \frac{\sigma_t(t)}{\varepsilon_0} \neq \frac{\sigma_{t_0}}{\varepsilon(t)} = \frac{1}{D(t)} \qquad (2\text{-}13)$$

as was discussed in the example mentioned earlier.

Other moduli and compliances can be defined and they are not, in general, equal to the functions defined above. It must be remembered that $G(t)$ and $E(t)$ can only be measured directly from constant strain experiments while $J(t)$ and $D(t)$ can only be measured directly from constant stress experiments. This fact is sometimes forgotten and can easily lead to considerable error.

C.   Dynamic Experiments

In addition to creep and stress relaxation experiments, another type of measurement is quite common. Here the stress or strain, instead of being a step function, is an oscillatory function with an angular frequency ω. Dynamic modulus values measured using such perturbations are functions of ω rather than time. The problem of putting dynamic experiments on a quantitative level is only slightly more difficult than is the case with step deformation experiments. In order to become familiar with this situation, we will begin by considering a simple experiment.

First suppose that a rod shaped sample of a completely elastic substance is fixed in a chuck and rotated at some frequency ω. Then

Figure 2-4. A Novel Dynamic Experiment. After Maxwell.[9]

by means of a suitable bearing device, we hang a weight W_0 from the end of the rod as shown in Figure 2-4. We know from experience that the rod will deform in the manner shown in Figure 2-5a (i.e., "straight down"). Formally, it may be said that stress and strain remain in phase regardless of the frequency and independently of the time elapsed since the start of the experiment.

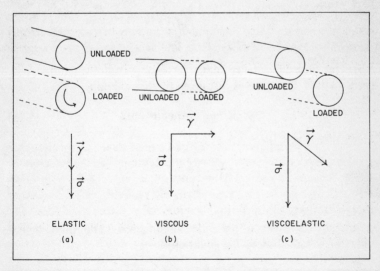

Figure 2-5. Actual Displacements of a Rod Subjected to the Dynamic Experiment Shown in Figure 2-4: (a) an Elastic Rod, (b) a Viscous Rod, (c) a Viscoelastic Rod.

A short analysis of the stress function will be helpful. In normal laboratory coordinates, we consider the force to be "downward" (stationary) and for the sample to be rotating at frequency ω. Considering the reference system to be the sample instead of the laboratory reveals, however, that this is the same as having the sample stationary and the weight rotating about the sample at frequency ω. This is just another view of the same experiment. At some time in the rotation of the weight all of the stress σ_0 is in the *down* direction. As the weight moves up through the cycle, the stress in the *down* direction decreases until, after 90° of rotation, the stress in this direction is zero. The stress continues to decrease and becomes $-\sigma_0$ at 180° from down, i.e., up, etc. The stress application in any direction can be treated as the projection of a vector of length σ_0 rotating at frequency ω on an axis in the direction of the stress being considered. The projection of a rotating vector on an axis yields a sine shaped function; the stress application in any direction of the sample is thus sinusoidal as a function of time. The situation is shown in Figure 2-6a.

For this elastic rod the modulus is *not* time-dependent. Thus we may write

$$\gamma(t) = \frac{\sigma(t)}{G} \tag{2-14}$$

where the time dependence comes exclusively from the nature of the perturbation. The stress is just given by the function

$$\sigma(t) = \sigma_0 \sin \omega t \tag{2-15}$$

which then yields the strain as

$$\gamma(t) = \frac{\sigma_0}{G} \sin \omega t \tag{2-16}$$

This behavior is shown in Figure 2-6b.

Secondly consider a perfectly viscous rod. The fundamental behavior is described by the equation

$$\frac{d\gamma}{dt} = \frac{\sigma(t)}{\eta} \tag{2-17}$$

(a) Stress in the down direction.

(b) Strain in the same direction for an
       elastic rod.

(c) Strain in the same direction for
       a viscous rod.

Figure 2-6. Stress and Strain in a Dynamic Experiment.

The strain rate, $d\gamma/dt$ responds linearly to the stress for a viscous body with a proportionality constant η, called the viscosity coefficient or just the viscosity.

In equation (2-17) it is clear that the rate of strain is a maximum when the stress is a maximum and a minimum when the stress is a minimum. Since a sine wave has its maximum rate of change when it passes through zero amplitude and its minimum rate of change when the amplitude is a maximum and since the stress in our experiment is sinusoidal, the stress and the strain must be 90° out of phase for a sample whose properties are defined by equation (2-17).

Mathematically, if $\sigma(t)$ is given by equation (2-15), equation (2-17) then becomes

$$\frac{d\gamma}{dt} = \frac{\sigma_0}{\eta} \sin \omega t \qquad (2\text{-}18)$$

which can be integrated to yield

$$\gamma(t) = \frac{\sigma_0}{\eta\omega} - \frac{\sigma_0}{\eta\omega}\cos\omega t \qquad (2\text{-}19)$$

This equation is plotted in Figure 2-6c and the configuration of a perfectly viscous rod subjected to these experimental conditions is that shown in Figure 2-5b.

Finally, the general case of a viscoelastic rod will result in a displacement somewhat between the two extremes, i.e., the strain will lag the stress by between 0° and 90° (Figure 2-5c).

It is not often that a dynamic experiment is carried out with the stress vector immobile in the laboratory coordinates and the sample rotating. Usually the sample is stationary in the laboratory and responds to an oscillatory stress or strain. Both of these situations are identical, however, except for the experimenter's point of view and from both of them he can measure the same physical properties.

Now consider a sample of viscoelastic material subjected to a sinusoidal shear stress. This is easily visualized using the technique of rotating vectors. In Figure 2-7 the magnitude of the vector **σ** represents the maximum of the stress applied to our sample. It rotates counterclockwise at a frequency ω. The actual stress on the sample is represented by the projection of this vector onto a suitable

(a)   (b)   (c)

Figure 2-7.   (a) Orientation of Stress and Strain Vectors in a Dynamic Experiment. (b) Stress Decomposed. (c) Strain Decomposed.

axis such as the y axis. Thus as $\boldsymbol{\sigma}$ rotates, its projection on the y axis is the sinusoidal stress experienced by the sample. As was mentioned previously, the strain will lag the stress by some amount, which is usually called the loss angle δ. In the rotating vector scheme, the strain is represented, as shown in Figure 2-7 by the vector $\boldsymbol{\gamma}$ which rotates at the same frequency as $\boldsymbol{\sigma}$ with a magnitude proportional to the maximum strain. The loss angle δ is the angle separating $\boldsymbol{\gamma}$ and $\boldsymbol{\sigma}$. The strain in the sample, likewise, is represented by the projection of the strain vector on the same axis as was used for the stress vector projection. It is obvious from the figure that the maximum in stress does not coincide with the maximum in strain.

The absolute shear modulus $|G|$ is defined as the magnitude of the stress vector divided by the magnitude of the strain vector; the absolute shear compliance $|J|$ is the reciprocal of this quantity.

Quite often it is convenient to separate the viscoelastic response into "in phase" and "out of phase" components. This is done in Figure 2-7. Here the projection of $\boldsymbol{\sigma}$ onto $\boldsymbol{\gamma}$ yields σ' the component of $\boldsymbol{\sigma}$ in phase with the strain while the projection of $\boldsymbol{\sigma}$ on the axis perpendicular to $\boldsymbol{\gamma}$ yields σ'' the component 90° out of phase with the strain. The in phase and out of phase shear moduli, G' and G'', are given as*

$$G' = \frac{\sigma'}{\gamma} \qquad G'' = \frac{\sigma''}{\gamma} \qquad (2\text{-}19a)$$

Similarly, the projection of $\boldsymbol{\gamma}$ onto $\boldsymbol{\sigma}$ yields γ' the strain component in phase with the stress; the projection of $\boldsymbol{\gamma}$ onto a vector axis perpendicular to $\boldsymbol{\sigma}$ yields γ'' the out of phase component of the strain. Here the shear compliance functions J' and J'' are defined as

$$J' = \frac{\gamma'}{\sigma} \qquad J'' = \frac{\gamma''}{\sigma} \qquad (2\text{-}20)$$

In Figure 2-8 the vectors are rearranged to make it clear that

$$\tan \delta = \frac{\sigma''}{\sigma'} = \frac{J''}{J'} = \frac{\gamma''}{\gamma'} = \frac{G''}{G'} \qquad (2\text{-}21)$$

where $\tan \delta$ is often called the loss tangent. Parameters with one

---

* Boldface type indicates vectors. Italic type indicates magnitudes.

σ	AC	σ''	CE
γ	AB	γ'	AF
σ'	AE	γ''	BF

Figure 2-8. Rearrangement of Figure 2-7.

prime are called storage functions and those with two primes loss functions. This has to do with the fact that in phase stress and strain results in elastically stored energy which is completely recoverable while one-fourth cycle separated stress and strain results in the dissipation of energy which is lost to the system (Problem 2-4).

In the above considerations, a sinusoidal shear stress was applied to the sample. It should be clear that a sinusoidal tensile stress is equally applicable and that definitions of $E'(\omega)$, $E''(\omega)$, $|D|$, D, etc., would yield results which would be completely analogous to the derived shear parameters.

The nomenclature of complex moduli and compliances is also often used. Here the out of phase component is made the imaginary part of a complex parameter; thus the complex shear modulus G^* and the complex shear compliance J^* are defined as

$$G^* = G' + iG''$$
$$J^* = J' - iJ''$$

(2-22)

The use of complex numbers to represent the functions has no particular significance except that by using the complex plane, one has a simple method for representing orthogonal vectors.

D. Boltzmann Superposition Principle

The Boltzmann superposition principle is one of the simplest but most powerful principles of polymer physics.[10] We have previously defined the shear creep compliance as relating the stress and strain in a creep experiment.*

$$\gamma(t) = \sigma_0 J(t) \qquad (2\text{-}23)$$

The stress σ_0 is applied instantaneously at time equal to zero. One might, however, imagine an experiment where the stress σ_1 is applied, not at $t = 0$, but at some other arbitrary time, perhaps u_1. Then equation (2-23) would become

$$\gamma(t) = \sigma_1 J(t - u_1) \qquad (2\text{-}24)$$

Consider now the application of two stresses σ_0 and σ_1 at the times $t = 0$ and $t = u_1$ respectively. The Boltzmann superposition principle asserts that the two stresses act independently and the resultant strains add linearly. This situation is shown in Figure 2-9. Thus

$$\gamma(t) = \sigma_0 J(t) + \sigma_1 J(t - u_1) \qquad (2\text{-}25)$$

or for a more general experiment consisting of discrete stresses $\sigma_1, \sigma_2, \ldots \sigma_n$ applied at times $t = u_1, u_2 \cdots u_n$

$$\gamma(t) = \sum_{i=1}^{n} \sigma_i J(t - u_i) \qquad (2\text{-}26)$$

The summation of the individual σ_i's would represent the total stress so that in considering a continuous stress application, $\sigma(t)$, the increment of applied stress is just the derivative of $\sigma(t)$. Replacing

---

* The subscripts t and s will be dropped from σ. J and G will be considered to result from σ_s. D and E will be considered to result from σ_t. All arguments presented concerning shear variables are completely applicable to tensile parameters also.

Figure 2-9. Linear Addition of Strains Applied Sequentially to a Sample.

the summation by an integration and remembering that u is the variable results in

$$\gamma(t) = \int_{-\infty}^{t} \frac{\partial \sigma(u)}{\partial u} J(t - u) \, du \qquad (2\text{-}27)$$

The limits of integration are taken as $-\infty$ since the complete stress history contributes to the observed strain and t since it is obvious that stresses applied after t, the time of observation of the strain, can have no effect on the observed strain.

In a completely analogous manner one may derive an expression relating the stress $\sigma(t)$ to the strain in a sample which has experienced some continuous strain history given by the function $\gamma(t)$

$$\sigma(t) = \int_{-\infty}^{t} \frac{\partial \gamma(u)}{\partial u} G(t - u) \, du \qquad (2\text{-}28)$$

Equations (2-27) and (2-28) are often given in an alternative form which we will now derive. Integrating equation (2-27) by parts

$$\int w\, dv = -\int v\, dw + \int d(wv) \qquad (2\text{-}29)$$

where

$$dv = \left(\frac{\partial \sigma(u)}{\partial u}\right) du \qquad w = J(t-u)$$

one obtains

$$\gamma(t) = J(t-u)\sigma(u)\Big|_{-\infty}^{t} - \int_{-\infty}^{t} \sigma(u)\frac{\partial J(t-u)}{\partial u}\, du \qquad (2\text{-}30)$$

We assume that $\sigma(-\infty)$ is equal to zero. Setting $t-u$ equal to a, a new variable, and observing new limits of integration due to this variable change gives

$$\gamma(t) = J(0)\sigma(t) + \int_{0}^{\infty} \sigma(t-a)\frac{\partial J(a)}{\partial a}\, da \qquad (2\text{-}31)$$

In an analogous manner, equation (2-28) becomes

$$\sigma(t) = G(0)\gamma(t) + \int_{0}^{\infty} \gamma(t-a)\frac{\partial G(a)}{\partial a}\, da \qquad (2\text{-}32)$$

As a specific example of the use of the Boltzmann principle, consider a material with a creep compliance given by the function

$$J(t) = J_r + t/\eta \qquad (a)$$

where J_r represents the recoverable deformation and η is the viscosity. Let us calculate the strain $\gamma(t)$ when this body is subjected to the linear stress function $\sigma(t)$ shown in part (a) of Figure 2-10; (i) at the time t_1 during the loading and (ii) at the time t_2 after the loading has ceased.

i The stress can be written as

$$-\infty \le t \le 0 \qquad \sigma(t) = 0$$
$$0 \le t \le t_1 \qquad \sigma(t) = kt$$

Making use of equation (2-27) one has

$$\gamma(t_1) = \int_{0}^{t_1} k\left[J_r + \frac{t_1}{\eta} - \frac{u}{\eta}\right] du \qquad (b)$$

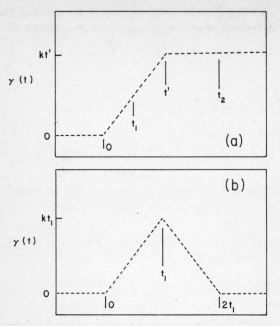

Figure 2-10. Strain Functions Used in Calculations.

and carrying out this simple integration leads to

$$\gamma(t_1) = kt_1 J_r + \frac{kt_1^2}{2\eta} \tag{c}$$

Remembering that kt_1 is just the total applied stress at the time t_1 yields

$$\gamma(t_1) = \sigma(t_1)\left[J_r + \frac{t_1}{2\eta}\right] \tag{d}$$

Equation (d) then gives an expression for the strain at time t_1 in a body whose creep compliance is given by equation (a) when it is subjected to a linear loading pattern starting at time zero. Note that t_1 must occur during the stressing period.

ii We will now calculate the strain after the stress addition has stopped. Again we can summarize the stressing history

$$-\infty \le t \le 0 \qquad \sigma(t) = 0$$
$$0 \le t \le t' \qquad \sigma(t) = kt$$
$$t' \le t \le t_2 \qquad \sigma(t) = kt'$$

Now equation (2-27) becomes

$$\gamma(t_2) = \int_0^{t'} k\left[J_r + \frac{t_2}{\eta} - \frac{u}{\eta}\right] du \qquad (e)$$

Again, this is a simple integration which gives the result that

$$\gamma(t_2) = \sigma(t')\left[J_r + \frac{t_2}{\eta} - \frac{t'}{2\eta}\right] \qquad (f)$$

It is interesting to carry out part (ii) of this example using the more widely used equation (2-31) instead of equation (2-27). One must use the new variable in the integral so that the strain history must be introduced in terms of this transformed variable rather than in terms of the normal laboratory time. Writing equation (2-31) for part (ii) of the example gives

$$\gamma(t_2) = J(0)\sigma(t_2) + \int_0^\infty \sigma(t_2 - a)\frac{\partial J(a)}{\partial a} da \qquad (g)$$

In laboratory time, there was no stress imposed between the time $t = -\infty$ and $t = 0$. In terms of the variable a, however, this corresponds to $a = \infty$ and $a = t_2$ for when $a = \infty$, $t_2 - a = -\infty$ and when $a = t_2$, $t_2 - a = 0$. This is a consequence of the variable change used to derive equation (2-31). Application of this equation without changing variables will necessarily lead to an incorrect result. Completing the stress summary in the usual way

$$-\infty \le t \le 0 \qquad \sigma(t) = 0 \qquad\qquad \infty \ge a \ge t_2$$
$$0 \le t \le t' \qquad \sigma(t) = kt \qquad\qquad t_2 \ge a \ge t_2 - t'$$
$$t' \le t \le t_2 \qquad \sigma(t) = kt' \qquad\qquad t_2 - t' \ge a \ge 0$$

From equation (a)

$$\frac{\partial J(a)}{\partial a} = \frac{1}{\eta} \qquad (h)$$

Substituting equation (h) and the strain history into equation (g) yields

$$\gamma(t_2) = J_r\sigma(t') + \int_0^{t_2-t'} \frac{k(t')}{\eta}\, da + \int_{t_2-t'}^{t_2} \frac{k}{\eta}(t_2 - a)\, da \qquad \text{(i)}$$

Integration and cancellation gives (f) as indeed it must.

Any of equations (2-27, -28, -31, -32) is completely sufficient as a statement of the Boltzmann superposition principle. Often in particular applications, however, it is more convenient to use one form rather than another.

E. The Relationship Between the Creep Compliance and the Stress Relaxation Modulus

One of the direct consequences of the Boltzmann superposition principle is that there is a relationship between the stress relaxation modulus and the creep compliance. We have already seen that, when dealing with time-independent functions, the compliance and the modulus are simply the reciprocals of each other. This simple relationship no longer holds in the time-dependent case. Fortunately, the Boltzmann superposition principle allows us to express moduli and compliances in terms of one another even in the time-dependent case. We will derive these relationships directly from equations (2-31) and (2-32) with the aid of LaPlace transforms. A short introduction or review of LaPlace transform techniques will be presented first.

Transform techniques in general are extremely powerful mathematical tools. The manner in which a transform operates is to take a problem in equation form from one space, where its solution is difficult, to another space where, it is hoped, the solution will be simpler. The solution in transform space is then transformed back into the original space to yield the answer to the problem. The LaPlace transform of a function $F(t)$ is denoted $f(p)$ or $L(F(t))$ and is defined as

$$f(p) \equiv L(F(t)) \equiv \int_0^\infty e^{-pt}F(t)\, dt \qquad (2\text{-}33)$$

It will be informative to derive several relationships among LaPlace transforms which will be used later. These relationships are to be found in any table of LaPlace transforms.

First consider the calculation of the LaPlace transform of the function

$$F(t) = at \tag{2-34}$$

where a is a constant.

Substitution into equation (2-33) gives

$$L(at) = \int_0^\infty e^{-pt} at \, dt \tag{2-35}$$

which upon integration yields

$$L(at) = \frac{a}{p^2} \tag{2-36}$$

Thus it is clear

$$L(t) = \frac{1}{p^2} \qquad L(aF(t)) = aL(F(t)) \tag{2-37}$$

Next consider the LaPlace transform of the function $F(t - a)$, where $F(x) = 0$ for $x < 0$. Again, substitution into equation (2-23) gives

$$L(F(t - a)) = \int_0^\infty e^{-pt} F(t - a) \, dt \tag{2-38}$$

Now letting $t - a = x$, one has

$$L(F(t - a)) = e^{-ap} \int_0^\infty e^{-xp} F(x) \, dx = e^{-ap} L(F(t)) \tag{2-39}$$

Lastly consider the transform of $F'(t)$. Proceeding as above

$$L(F'(t)) = \int_0^\infty e^{-pt} F'(t) \, dt \tag{2-40}$$

Integration by parts yields

$$L(F'(t)) = e^{-pt} F(t)|_0^\infty + p \int_0^\infty e^{-pt} F(t) \, dt \tag{2-41}$$

The second term is just the definition of the LaPlace transform of $F(t)$; evaluation of the first term at the limits of integration gives

$$L(F'(t)) = -F(0) + pL(F(t)) \tag{2-42}$$

It is clear that this expression is very powerful allowing one to find the transform of the derivative of a function from a knowledge of the function itself; it is not necessary to know the derivative of the function explicitly.

One additional result will be needed which will be derived in Appendix 2 of this chapter. This is Borel's theorem which states

$$L\left(\int_0^t F_1(t - \tau) \, F_2(\tau) \, d\tau\right) = L(F_1(t)) \, L(F_2(t)) \tag{2-43}$$

Using these results we can now derive the relationship between the creep compliance and the stress relaxation modulus.

The LaPlace transform of equation (2-31) yields

$$L(\gamma(t)) = J(0) \, L(\sigma(t)) + \int_0^\infty e^{-pt} \int_0^\infty \sigma(t - a) \frac{\partial J(a)}{\partial a} \, da \, dt \tag{2-44}$$

or

$$L(\gamma(t)) = J(0) \, L(\sigma(t)) + \int_0^\infty \frac{\partial J(a)}{\partial a} \int_0^\infty e^{-pt} \sigma(t - a) \, dt \, da \tag{2-45}$$

Making use of the result derived in equation (2-43) gives

$$L(\gamma(t)) = J(0) \, L(\sigma(t)) + \left[\int_0^\infty e^{-ap} \frac{\partial J(a)}{\partial a} \, da \right] L(\sigma(t)) \tag{2-46}$$

The term enclosed in brackets, however, is nothing except the LaPlace transform of the derivative of $J(t)$. Thus we may apply the result obtained in equation (2-45) to get

$$L(\gamma(t)) = J(0) \, L(\sigma(t)) + L(\sigma(t)) \, [pL(J(t)) - J(0)] = pL(\sigma(t)) \, L(J(t)) \tag{2-47}$$

Next, transform equation (2-32) into LaPlace space in the same manner to obtain

$$L(\sigma(t)) = G(0) \, L(\gamma(t)) + L(\gamma(t)) \, [pL(G(t)) - G(0)]$$
$$= pL(\gamma(t)) \, L(G(t)) \tag{2-48}$$

Equations (2-47) and (2-48) give

$$\frac{1}{p^2} = L[G(t)]L[J(t)] \qquad (2\text{-}49)$$

This is the solution of the problem in transform space. We have a direct relationship between the transforms of the compliance and the modulus. This solution must now be returned to real space. Making use of Borel's theorem, equation (2-43) and the result derived in equation (2-33) gives the final result

$$t = \int_0^t G(\tau)\, J(t - \tau)\, d\tau \qquad (2\text{-}50)$$

This is the convolution integral which is the relationship between the creep compliance and the stress relaxation modulus. It is exact and depends only on the applicability of the Boltzmann superposition principle.

These equations are used to convert a modulus to a compliance in problem 6 of this chapter. The results of this calculation are depicted in Figure 2-11 (see reference 11).

Figure 2-11. Relationship Between a Creep Compliance and a Stress Relaxation Modulus (Special Case).

F. The Relationship Between Static and Dynamic Properties

Still another relationship between experimental parameters is a direct consequence of the Boltzmann superposition principle. We will derive the equations relating the tensile stress relaxation modulus $E(t)$ to the in-phase and out-of-phase dynamic tensile moduli $E'(\omega)$ and $E''(\omega)$ starting from equation (2-24) expressed in tensile quantities. Equation (2-28) may be written as

$$\sigma(t) = \int_{-\infty}^{t} \frac{\partial \gamma(u)}{\partial u} E(t - u) \, du \qquad (2\text{-}51)$$

A simple variable change setting $s = t - u$ yields

$$\sigma(t) = -\int_{0}^{\infty} \frac{\partial \gamma(t - s)}{\partial s} E(s) \, ds \qquad (2\text{-}52)$$

Consider the application of a sinusoidal strain which may be represented by

$$\gamma(t) = \gamma_0 e^{i\omega t} = \gamma_0(\cos \omega t + i \sin \omega t) \qquad (2\text{-}53)$$

where γ_0 is the maximum amplitude of the strain. Substitution of equation (2-53) into equation (2-52) and simplification gives

$$\frac{\sigma(t)}{\gamma(t)} = \int_{0}^{\infty} i\omega e^{-i\omega s} E(s) \, ds \qquad (2\text{-}54)$$

By expanding the term e^{-iws}, as shown in equation (2-53), one obtains

$$E^*(\omega) = \int_{0}^{\infty} \omega(\sin \omega s) E(s) \, ds + i \int_{0}^{\infty} \omega(\cos \omega s) E(s) \, ds \quad (2\text{-}55)$$

Here we have observed that the ratio of the tensile stress to the tensile strain in an experiment where the perturbing function may be given a complex sinusoidal representation is the complex tensile modulus $E^*(\omega)$. From this expression, one immediately realizes that either of the two dynamic modulus functions may be calculated from the tensile relaxation modulus as follows:

$$E'(\omega) = \omega \int_0^\infty \sin \omega s \, E(s) \, ds$$

$$E''(\omega) = \omega \int_0^\infty \cos \omega s \, E(s) \, ds$$

(2-56)

This is a specific example illustrative of the method by which one may relate static and dynamic properties. Listings of other results may be found in texts such as Ferry.[3]

It is apparent that equation (2-56) embodies a Fourier sine and cosine transformation of $E(t)$; thus normal Fourier transform[12] methods permit the inversion of these relations to give the static modulus as a function of the dynamic properties.

Also in this section we have tacitly assumed that $E(t)$ approaches zero at long times. If this is not true, as is the case with a crosslinked polymer, one may introduce the function $E(t) - E_e$ where E_e represents the equilibrium tensile modulus. It is clear that this difference does indeed become zero at long times.

Appendix 1:   The Relationship Between Tensile and Shear Parameters

$$\mu = \frac{1}{2}\left[1 - \frac{1}{V}\left(\frac{dV}{d\varepsilon}\right)\right]$$

(a)

Consider the application of a small tensile stress to a sample as shown in Figure 2-1a. The volume will change by some amount ΔV under this perturbation and the tensile strain will change from 0 to ε. Thus equation (a) becomes

$$\mu = \frac{1}{2}\left(1 - \frac{1}{\varepsilon}\frac{\Delta V}{V}\right)$$

(b)

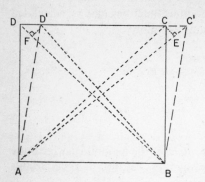

Figure 2-12. Body Deformed in Shear.

Since ε is small, however, the fractional volume change is just $\varepsilon - 2d$ where d is $|\Delta A/A|$ or $|\Delta B/B|$ (A and B both decrease). This result obtains since for the figure shown

$$\frac{\Delta V}{V} \approx \frac{AB\,\Delta C + AC\,\Delta B + BC\,\Delta A}{ABC}$$

$$\approx \frac{\Delta C}{C} + \frac{\Delta B}{B} + \frac{\Delta A}{A} = \varepsilon - 2d \tag{c}$$

Substituting equation (c) into equation (b) yields the simple result

$$\mu = \frac{d}{\varepsilon} \tag{d}$$

which is an alternate definition of the Poisson ratio. From here on we will use the proof outlined in reference 1. First we must show that a shear strain γ can be replaced by two tensile strains each of magnitude $\gamma/2$.

In Figure 2-12 we show a cross section of a cube $ABCD$ which is sheared as shown. In addition several diagonals have been drawn. Here the shear strain γ is clearly $\tan\theta$ and equal to CC'/CB.

Construct the line CE which is perpendicular to the rhombus diagonal AC'. Since the strain is small, $\angle CC'E$ is close to $\pi/4$ and $C'E = CC'/\sqrt{2}$. Similarly $AE \approx AC$. Now define the tensile strain along the diagonal AC' as

$$\varepsilon = \frac{C'E}{AE} = \frac{CC'/\sqrt{2}}{AC} = \frac{CC'/\sqrt{2}}{CB\sqrt{2}}$$

or   (e)

$$\varepsilon = \frac{1}{2}\frac{CC'}{CB} = \frac{\gamma}{2}$$

Similarly we could show that a negative strain of magnitude $\gamma/2$ results from consideration of the motion of corner D. Thus a shear strain can be decomposed into two perpendicular tensile strains, each half the magnitude of the tensile strain, of opposite sign and both oriented 45° to the shear strain.

Thus let linear stresses σ_t and $-\sigma_t$ which are perpendicular to each other act on a sample to generate a shear strain which is oriented 45° away from both of these forces. In the direction of $+\sigma_t$, the tensile strain ε is just

$$\varepsilon = \frac{\sigma_t}{E} + \mu\frac{\sigma_t}{E} \tag{f}$$

The first term on the right hand side arises from the fact that the sample elongates in the direction of the $+\sigma_t$ application because of the $+\sigma_t$ application; and simultaneously, it elongates even more due to the application of the $-\sigma_t$ stress perpendiculiar to the $+\sigma_t$ direction. From equations (d) and (f)

$$\mu\frac{\sigma_t}{E} = \frac{d}{\varepsilon}\frac{\sigma_t}{E} = \frac{d}{\varepsilon}\varepsilon = d \tag{g}$$

Since the tensile strain is just $\frac{1}{2}$ the magnitude of the shear strain, equation (f) becomes

$$\gamma = \frac{2\sigma_t}{E}(1 + \mu) \tag{h}$$

Finally we must show that a tensile stress of magnitude σ_t acting in the x direction and a tensile stress of magnitude σ_t acting in the Z direction is equivalent to a shear stress σ_s of magnitude σ_t acting in a direction 45° to each tensile stress. To do this, consider Figure 2-13 where these stresses are shown. This is just a cross section of a three-dimensional cube. The actual magnitude of the force on the

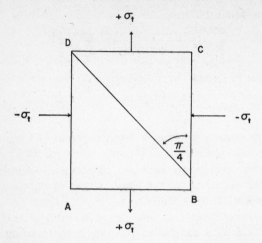

Figure 2-13.　Shear as a Result of Tensile Stresses.

surface (surface area S) represented by the line BC is just $S\sigma_t$ and is numerically equal to the magnitude of the forces on the plane represented by the line DC. Now construct the line DE as shown. The resultant of the forces mentioned above is just $(\sqrt{2}S\sigma_t)$ and directed *along* the new plane represented by the line DB. The shear stress along this plane then is just obtained by dividing the force along the plane by the area it is acting upon to give (area of the plane represented by DB is $\sqrt{2}S$)

$$\sigma_s = \frac{\sqrt{2}\,S\sigma_t}{\sqrt{2}\,S} = \sigma_t \tag{i}$$

By definition, however,

$$\sigma_s = G\gamma \tag{j}$$

Nevertheless, according to equations (h) and (i)

$$\sigma_t = G\frac{2\sigma_t}{E}(1 + \mu)$$

which can be rearranged to give

$$E = 2G(1 + \mu) \tag{k}$$

Appendix 2:   Borel's Theorem

Consider the two functions $f_1(t)$ and $f_2(t)$ whose LaPlace transforms are given as $F_1(p)$ and $F_2(p)$ respectively. Then the product

$$F_1(p)F_2(p) = \int_0^\infty f_1(T)e^{-pT} \int_0^\infty e^{-pt} f_2(t) \, dt \, dT \tag{a}$$

If

$$f_2(t) = 0 \quad \text{for} \quad t < 0 \tag{b}$$

however, application of equation (2-39) yields

$$F_1(p)F_2(p) = \int_0^\infty f_1(T) \int_0^\infty e^{-pt} f_2(t - T) \, dt \, dT \tag{c}$$

Upon rearrangement we get

$$F_1(p)F_2(p) = \int_0^\infty e^{-pt} \int_0^\infty f_1(T)f_2(t - T) \, dT \, dt \tag{d}$$

and remembering condition (b) yields

$$F_1(p)F_2(p) = \int_0^t f_1(T)f_2(t - T) \, dT$$

Problems

1.   Calculate the weight needed to bring about a shear deformation, ΔX, of 0.40 cm, of a viscoelastic cube 2 cm on a side after 10^{-4}, 10^{-2}, 10^0, 10^4, and 10^6 seconds. The time-dependent shear compliance is

$$J(t) = \left(10^{-10} + \frac{t}{10^8}\right)\frac{\text{cm}^2}{\text{dyne}}$$

2.   Make a table expressing J', J'', $\tan \delta$, $|J|^2$, $|G|^2$, G', and G'' in terms of (a) G' and G'', (b) J' and J'', (c) $|G|$ and $\tan \delta$, and (d) $|J|$ and $\tan \delta$.

3.   Calculate the stress at time t longer than $2t_1$ in a body with a relaxation modulus given by

$$E(t) = E_1 e^{-t/\tau_1} + E_2 e^{-t/\tau_2}$$

when the body is subjected to the strain history depicted in Figure 2-10b.

4. By considering a sinusoidal stress applied to a viscoelastic body, show that in phase strain results in conservation of energy (work) while out of phase strain results in energy dissipation.

5. Show that

$$J(t) = \frac{\sin m\pi}{m\pi} \frac{1}{G(t)}$$

if $\log J(t) = \log A + m \log t$. (Note $\Gamma(n) \Gamma(1 - n) = \pi/\sin n\pi$.)

6. Calculate the creep compliance for a body whose stress relaxation modulus is given by

$$G(t) = G_0 e^{-t/\tau}$$

References

1. R. B. Lindsay, *Physical Mechanics*, Van Nostrand, New York, 1933, pp. 317–324.

2. L. Malvern, *Introduction to the Mechanics of a Continuous Medium*, Prentice-Hall, New York, 1969.

3. J. D. Ferry, *Viscoelastic Properties of Polymers*, Wiley, New York, 1970.

4. A. V. Tobolsky, *Properties and Structure of Polymers*, Wiley, New York, 1960.

5. F. Bueche, *Physical Properties of Polymers*, Interscience, New York, 1962.

6. B. Gross, *Mathematical Structure of the Theories of Viscoelasticity*, Hermann & Cie, Paris, 1953.

7. F. R. Eirich, Ed., Rheology, *Theory and Application*, Academic Press, New York, 1956.

8. D. J. Plazek, *J. Polymer Sci.*, A-2 **6**, 621 (1968).

9. B. Maxwell, *J. Polymer Sci.*, **20**, 551 (1956).

10. H. Leaderman, *Elastic and Creep Properties of Filamentous Materials*, The Textile Foundation, Washington, D.C., 1943.

11. I. L. Hopkins and R. W. Hamming, *J. Appl. Phys.*, **28**, 906 (1957).

12. S. Goldman, *Transformation Calculus and Electrical Transients*, Prentice-Hall, New York, 1949.

3 Time–
Temperature
Correspondence

Having examined some of the basic manifestations of the phenomenological aspects of viscoelasticity, we now shift our emphasis to interpretations of this behavior on a molecular scale. The concepts of the previous chapter are strictly independent of the existence of molecules; the results of those concepts fall into the realm of continuum mechanics. In this chapter we will observe the results of generalized experiments and then interpret specific behavioral patterns in terms of the accepted molecular mechanisms. Furthermore, the basis of ideas concerning time-temperature correspondence will be introduced.

A.   Four Regions of Viscoelastic Behavior

Once again, begin by considering an experiment. A polymer sample of unit cross sectional area is subjected to an instantaneous tensile strain which is thereafter maintained constant. The stress is monitored as a function of time and the tensile stress relaxation modulus is obtained using equation (2-11). Let t in equation (2-11) be any arbitrary time, perhaps 10 seconds. Next

37

the stress is removed allowing the sample to relax and the temperature is changed. The same experiment is carried out yielding E (10 seconds) at this new temperature. The experiment is repeated at many temperatures to yield the "ten-second tensile relaxation modulus" as a function of temperature.

The types of behavior most often observed are shown in Figures 3-1 to 3-3. (We have selected tensile relaxation as an example and other types of deformation would have been equally suitable.)

Figure 3-1 shows idealized modulus-temperature curves for typical linear and crosslinked amorphous polymers. In this plot four regions of viscoelastic behavior are defined. At low temperatures where the modulus is higher than 10^{10} dynes/cm², the polymer is hard and brittle; this is the glassy region. The glassy modulus, E_1, a slowly decreasing function of temperature, is a useful parameter to use in characterizing polymeric behavior. In this glassy region, thermal energy is insufficient to surmount the potential barriers for rotational and translational motions of segments of the polymer molecules. The chain segments are essentially "frozen" in fixed positions on the sites of a disordered quasi-lattice with their segments vibrating around these fixed positions much like low molecular

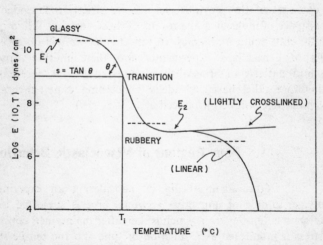

Figure 3-1. Idealized Modulus-Temperature Curve Showing Various Regions of Viscoelastic Behavior.

weight molecules in a molecular crystal. With increasing temperature, the amplitude of vibrational motion becomes greater and eventually the thermal energy becomes roughly comparable to the potential energy barriers to segment rotation and translation. In this temperature region the polymer is at the glass transition temperature where short-range diffusional motions begin. Segments are free to "jump" from one lattice site to another; the brittle glass becomes a resilient leather.

This phenomenon is accompanied by a catastrophic decrease in the modulus of several decades as indicated in Figure 3-1. The breadth of this transition region ranges from 5 to more than 20 degrees centigrade, depending on the nature of the polymer in question. The transition region may be characterized by two parameters. The temperature at which the 10-second modulus reaches 10^9 dynes/cm^2 is termed the inflection temperature, T_i, and the negative slope of the curve (tan θ) at this point is called s. Values of these characteristic viscoelastic parameters for several amorphous polymers are listed in Table 3-1.

As the temperature is further increased, the modulus again reaches a plateau region. This rubbery plateau is characterized by the modulus E_2 as shown in Figure 3-1. In this temperature interval, the short range diffusional motions of the polymer segments which initially gave rise to the glass transition occur very much faster than our measurement time of 10 seconds. On the other hand, the long range cooperative motion of chains which would result in translational motions of complete molecules is still greatly restricted by the presence of strong local interactions between neighboring chains. In the case of the crosslinked material, these interactions consist of primary chemical bonds. In the linear polymer, they are known as entanglements and their precise nature is not clear. In any case, in the rubbery plateau region, segments of chains reorient relative to each other but large scale translational motion does not occur.

The viscoelastic responses of linear and crosslinked polymers through the rubbery plateau region are essentially identical. As the temperature is further increased, however, differences between these two categories of polymers become evident as shown in Figure 3-1. First consider a crosslinked network. As temperature is increased,

TABLE 3-1. Characteristic Viscoelastic Parameters for Selected Polymers

Polymer	$3G_1$ (dynes/cm^2)	$3G_2$ (dynes/cm)	T_i (°C)	s (°C^{-1})	n
Crystalline Polymers					
Polyethylene ($M_w = 2.6 \times 10^5$)	3×10^{10}	3×10^7	75	0.01	—
Polyvinyl chloride	4	2	78	0.25	0.60
Amorphous Linear Polymers					
Polyisobutylene	3.5	1	−62	0.15	0.75
Natural rubber (unvulcd.)	2.5	4	−67	0.2	1.25
Polystyrene	2	0.5	101	0.2	0.85
Polymethyl acrylate	3	1.5	16	0.2	—
Polymethyl methacrylate	1.5	2	107	0.15	0.55
Polybutyl acrylate	1.5	0.5	−53	0.2	—
Polybutyl methacyrlate	1	1	31	0.15	—
cis-Polybutadiene	2	1	−106	0.2	—
Polyethylene tetrasulfide	2	4	−24	0.15	0.65
Atactic polypropylene	2	2.5	−16	0.2	—
Polyacenaphthalene	—	—	264	0.1	0.75
Polyethylhexyl acrylate	—	0.5	−70	0.2	—
Selenium	6	—	45	0.15	1.0
Sulfur	—	—	−28	0.2	—
Bisphenol A polycarbonate ($M_w = 4 \times 10^5$)	1.5	5	150	0.3	1.1
Polyethyl ether disulfide	0.5	2	−53	0.25	0.6
Polyethyl formal disulfide	1	0.5	−58	0.35	—
Amorphous Lightly Crosslinked Polymers					
Natural rubber (vulcd.)	3.5	4	−57	0.2	1.1
Polytetrahydrofuran	3	10	−73	0.15	—
Poly-2-hydroxylethyl methacrylate	3	2.5	96	0.1	0.55
Poly-t-butylaminoethyl methacrylate	2.5	1	41	0.1	—
Polyethyl methacrylate	2.5	2.5	77	0.1	—
Poly-n-propyl methacrylate	2	2	56	0.1	0.50
Polymethoxyethyl methacrylate	2	1.5	23	0.1	—
Polydiethylaminoethyl methacrylate	2	0.5	26	0.1	—

the crosslinks consisting of primary chemical bonds remain intact, preventing the chains from translating relative to one another. Thus, although the modulus will change slightly with temperature in the rubbery plateau region of a crosslinked polymer (Chapter 6), this is a small effect compared to changes like those exhibited during the glass transition. Thus, to a first approximation, the modulus will remain constant for a crosslinked rubber up to temperatures where chemical degradation begins to occur.

The situation is quite different for a linear polymer. In this case, increasing temperature causes molecular motions to become more and more large-scale until eventually whole polymer molecules begin to translate. When the temperature is high enough, local chain interactions are no longer of sufficiently high energy to prevent molecular flow. During the 10-second test, the molecules will slip by one another, releasing much of the local strain and the polymer sample will exhibit a correspondingly low stress value. Since the molecules are translating relative to one another in this region, it should be expected that a considerable amount of flow will result from experiments carried out under these conditions. When the strain is released in our stress relaxation experiment, the sample will not recover to its former length, but will relax to a new equilibrium state having a greater length. Thus, during the experiment, the sample has undergone flow and this area of the 10-second modulus temperature curve is therefore called the flow region.

If temperature is increased still further, barring chemical reaction, the sample will become a viscous liquid. Our simple experimental setup will no longer be applicable at high temperatures where the sample will not support its own weight. The modulus will, however, continue to decrease.

With this background, it will be easy to proceed with the analysis of the modulus-temperature behavior of more complicated systems.

First consider the effect of different molecular weight in linear amorphous systems. Figure 3-2 shows the relaxation behavior of three polystyrene samples. Sample (a) has a very low molecular weight in the neighborhood of 10,000, while samples (b) and (c) have molecular weights of 1.4×10^5 and 2.17×10^5 respectively. First consider samples (b) and (c). At temperatures below 120°C, both materials are identical. There is no differentiation between

Figure 3-2. Ten-Second Modulus Versus Temperature Curve for Polystyrene of Various Molecular Weights.

the two samples since short range segmental motions are determining the behavior. Such motions are essentially independent of molecular weight at high molecular weights. At higher temperatures, however, where translational motions of the complete molecules determine the relaxation behavior of the sample, it is most reasonable to expect longer molecules to maintain a pseudo-network to higher temperatures. This is indeed observed as shown in Figure 3-2. Sample (a), however, appears to behave anomalously. We notice that the initial decay of the glassy modulus occurs at lower temperatures than is the case with the high molecular weight polymers and, moreover, there is no rubbery plateau. The behavior of sample (a) in the glass transition region may be qualitatively accounted for on the basis of "free volume" considerations as follows (more detail is given in Chapter 4). Since a linear polymer chain has two ends, decreasing the length of the chains leads to an increase in the concentration of ends in the sample. However, since a chain middle is attached to other segments on both sides, it is less mobile than a chain end which is attached on only one side. Thus a polymer sample which has a greater concentration of chain ends than another comparable sample will exhibit greater chain mobility at a given temperature. In other words, a decrease in molecular weight leads to a decrease in glass transition temperature. The absence of a rubbery plateau is due to the fact that the short molecules present in this low molecular weight sample are not long enough to form an "entanglement" network which is necessary for the observation of a rubbery state. (There appears to be a critical molecular weight necessary for an entanglement network to form.)

The above considerations must be modified rather extensively for the case of a crystalline polymer. Figure 3-3 shows the 10-second modulus-temperature curve for two typical crystalline polymers, polyethylene (PE) and polyvinyl chloride (PVC), contrasted with atactic polystyrene (PS), a typical noncrystalline polymer.

It should be emphasized that a crystalline polymer is not totally crystalline but partly crystalline and partly amorphous. Thus one expects to see behavior characteristic of the crystalline regions superposed upon behavior characteristic of the amorphous regions. This superposition is not necessarily a linear one but may rather be a complex coupling of the response of each network. Furthermore, the degree of crystallinity as well as the crystallite size is not a unique feature of any polymer system. Both of these properties are determined by prior thermal history.

The general characteristics attributed to the amorphous network are not greatly modified in crystalline polymers. There is still a glass to rubber transition, a rubbery plateau, and finally, in

Figure 3-3. Comparison of Ten-Second Modulus Versus Temperature Curve for Polystyrene, Polyvinyl Chloride and Polyethylene.

noncrosslinked systems, a flow region. This amorphous network, however, couples with the crystalline portion of the polymer. The main transition exhibited by the crystalline part of the polymer is a melting from an ordered crystal to a liquid like, disordered state. The temperature of this melting is usually called T_m. With these thoughts in mind, one may now analyze the curves in Figure 3-3. Consider polyethylene. T_m is about 125°C. While the temperature of the glass transition is not yet known with certainty, it seems clear that the glass transition temperature is low. At temperatures between the glass transition temperature and T_m, the polymer exhibits a very .high modulus and experimentally polyethylene is also noted to have a very high impact strength in this region. Molecularly, this is to be expected since, in polyethylene, a large portion of the polymer is crystalline. Therefore, in this region one is observing the relaxation of a small amount of material held together by numerous hard crystallites. Here a high modulus should be expected. As T_m is approached, changes in the crystalline superstructure and in the degree of crystallinity result in a rapid decrease of the modulus. At the melting point, the modulus drops sharply to the rubbery modulus of polyethylene. With the crystallites gone, we are left with a normal amorphous polymer which will exhibit behavior character-istic of amorphous polymers. This, indeed, is so for polyethylene where a rubbery plateau and eventually a flow region are observed at temperatures above T_m.

The situation is somewhat similar in the case of polyvinyl chloride (PVC). PVC has a lower degree of crystallinity than polyethylene so that the viscoelastic response of PVC might be expected to approximate more closely that of an amorphous polymer than does polyethylene. Figure 3-3 indicates that this is indeed the case. Specifically, the modulus drop in PVC at T_g is considerably greater than that in polyethylene although the rubbery plateau modulus is higher than that of amorphous polymers and the rubbery plateau extends to the melting point of the PVC crystallites, about 180°C. In a polymer of low crystallinity like PVC, the rubbery plateau can be well explained by the postulate that the crystallites act as both a filler and as crosslinks.

Once again, it should be pointed out that the exact shape of the modulus-temperature curve of a crystalline polymer depends on the thermal history of the sample, particularly on the rate of cooling from the melt and annealing treatment. Two crystalline polymers are mechanically "equivalent" for practical purposes, if they have the same value of T_g, T_m, chain length, percent crystallinity, and crystalline structure.

The incorporation of soluble diluents (plasticization) into polymers alters the viscoelastic properties to a greater or lesser extent depending on the exact polymer diluent system and the amount of diluent present. One of the most important polymers used in the plasticized state is polyvinyl chloride.

Figure 3-4 shows that PVC plasticized with 30 weight percent dioctyl phthalate differs significantly from pure PVC in viscoelastic behavior. Although E_1 and E_2 appear to be essentially independent of the presence of diluents, other characteristic parameters show interesting effects. T_i, the inflection temperature, for example, is depressed, while the transition region is broadened. It was pointed out previously that rubbery flow commences in PVC when T_m is exceeded. Since T_m is depressed by 150°C due to the presence of diluent, the flow region begins at a lower temperature than for unplasticized PVC. If we regard diluent as part of the polymer system, then the occurrence of the flow region at lower temperatures

Figure 3-4. The Effect of Plasticization on a Modulus Temperature Curve. Reprinted from *Plasticization and Plasticizer Processes*, *Advances in Chemistry Series*, **98**, 1181 (1965) by permission of the American Chemical Society.

might be interpreted as being due to the lowering of the number average molecular weight.

Since the plasticizer molecules are small, the argument used to rationalize the low glass transition temperature of the 10,000 molecular weight polystyrene sample shown in Figure 3-2 seems applicable also to rationalize the decrease in the temperature of the glass to rubber transition in the plasticized system.

B.   Time–Temperature Superposition

Remembering that the modulus is a function of time as well as temperature leads one to wonder about the feasibility of measuring the modulus as a function of time at constant temperature instead of, as we have done throughout the first part of this chapter, measuring the modulus as a function of temperature at constant time. In principle the complete modulus versus time behavior of any polymer at any temperature can be measured. The experiment is straightforward as described in Chapter 2. A number of convenient approaches are available. The results of such an experiment are shown on the left hand side of Figure 3-5 where tensile stress relaxation has been chosen as the experiment. Clearly only a small

Figure 3-5. Preparation of a Stress Relaxation Master Curve from Experimentally Measured Modulus-Time Curves at Various Temperatures.

range of viscoelastic response manifests itself during an experimentally accessible time range of perhaps 10 seconds to 1 hour. Consider, for example, the experiment at temperature T_2. We observe that the modulus decays by a factor of about 10 in this period of time, a decrease in modulus far short of the five orders of magnitude which we observed in the modulus temperature experiments. If one were to carry out the experiment at a different temperature, different behavior would result; again this is shown on the left hand side of Figure 3-5. Clearly then, it is not feasible to directly measure the complete behavior of the modulus as a function of time at constant temperature using methods such as those described in Chapter 2.

Experimentalists have found a shifting procedure which enables one to construct a "master curve" (complete modulus-time behavior at a constant temperature) in spite of this limitation. This technique is based on the principle of time-temperature correspondence.

Consider, now, the curve measured at T_1 and terminating at about 1000 seconds. The time-temperature correspondence principle states that there are two methods to use in order to determine the polymer's behavior at longer times. First, one may directly measure the response at longer times. This technique rapidly becomes prohibitively time-consuming since the change is so slow (Figure 3-5 is plotted on a log scale). Secondly, one may increase the temperature to T_2, for example, and then carry out the relaxation experiment on an experimentally accessible time scale again. Now one observes that shifting curve T_2 horizontally to the right will result in an exact superposition of the curves measured at T_1 and T_2 in the areas where the modulus values overlap and an extension of the curve measured at T_2 to modulus values lower than those measured at temperature T_1. The time-temperature correspondence principle states that this extension is identical to that which would be measured at long times at the temperature T_1. Thus one effectively has a measure of the complete modulus-time behavior by applying the time-temperature correspondence principle to experimental measurements of polymer relaxations carried out on experimentally accessible time scales.

Mathematically these ideas may be expressed as

$$E(T,t) = E(T_2, t/a_T) \tag{3-1}$$

where the effect of changing temperature is the same as applying a multiplicative factor to the time scale, i.e., an additive factor to the log time scale.

Although this discussion of the time-temperature correspondence principle is essentially correct, one additional correction is necessary. We have, by shifting horizontally, compensated for a change in the time scale brought about by changing temperature. There is also, however, an inherent change in the modulus brought about by a change in temperature. In Chapter 6 it will be shown that the modulus of a rubbery network is directly proportional to T, the absolute temperature. Thus, in applying a reducing procedure to make a master curve for individual relaxation experiments, not only must one take into account the time scale shift, one must also consider that there will be a slight vertical shift due to the temperature variation. Similarly, since the volume of a polymer is a function of temperature and the modulus, being defined per unit cross sectional area, will obviously vary with the amount of matter contained in unit volume, a corresponding correction must be made so that changing mass per unit volume as a function of temperature is accounted for. The density is obviously the parameter which must be used. These considerations lead one to write[4]

$$\frac{E(T_1,t)}{\rho(T_1)T_1} = \frac{E(T_2,t/a_T)}{\rho(T_2)T_2} \tag{3-2}$$

Division by the temperature corrects for the changes in modulus due to the inherent dependence of modulus on temperature while division by the density corrects for the changing number of chains per unit volume with temperature variation.

When constructing a master curve, one arbitrarily picks a reference temperature, T_0. The modulus at any time t, which one would observe at the temperature T_0 in terms of the experimentally observed modulus values at different temperatures T, is therefore given as

$$E(T_0,t) = \frac{\rho(T_0)T_0}{\rho(T)T} E(T,t/a_T) \tag{3-3}$$

where a_T is to be discussed more fully. If one is considering compliance functions, equation (3-2) takes the form, using $J(t)$ as an example,

$$\rho(T_1)\, T_1 J(T_1,t) = \rho(T_2)\, T_2 J(T_2,t/a_T) \qquad (3\text{-}4)$$

and

$$J(T_0,t) = \frac{\rho(T)T}{\rho(T_0)T_0}\, J(T,t/a_T) \qquad (3\text{-}5)$$

C. Master Curves

We may now consider the preparation of a master curve from the data in Figure 3-5. First, let us arbitrarily pick T_3 as the reference temperature. With a knowledge of the density at all the temperatures T_i, one applies the vertical correction factor stated in equation (3-3) to all the curves. For T_3 the factor $\rho(T_0)T_0/\rho(T)T$ is unity resulting in no shift. This must be so since T_3 has been chosen as the reference. At the other temperatures, however, the correction factor will not in general be unity. Let us now consider that these corrections have been made to the experimental curves on the left hand side of Figure 3-5. The curve at T_3 is reproduced on the right hand side of the figure. Next, the curve at T_2 is shifted to the left giving rise to the dotted extension of the $E(T_3,t)$ curve. This process is repeated until the complete master curve is formed. Again this procedure is mathematically described by equation (3-3). The a_T's are functions of temperature and are known as the shift factors. The subscript T indicates that the shift factors are taken relative to some standard temperature. In our example, all of the curves at temperatures higher than T_3 are shifted to the right, those at T_1 and T_2 are shifted to the left. It should be clear that any temperature might have been chosen as the reference temperature. If T_4 had been chosen, for example, some of the shift factors would have been larger than 1.0—those at lower temperatures—while some of the shift factors would have been less than 1.0—those at temperatures higher than T_4.

D. The WLF Equation

It is now common practice to reduce most relaxation or creep data to the temperature T_g; thus, the reference temperature

Figure 3-6. The WLF Equation for Polystyrene Using Data from Table 3-2.

is picked as the glass transition temperature measured by some slow technique such as dilatometry. The function $\log a_T$ for this choice of reference temperature is shown in Figure 3-6. Here again, at the reference temperature a_T is 1.0 and $\log a_T$ is 0. All amorphous polymers exhibit similar behavior on such a plot. The equation of this curve is[7]

$$\log a_T = \frac{-C_1(T - T_g)}{C_2 + T - T_g} \tag{3-6}$$

The equation is called the WLF equation (Williams, Landel, and Ferry) after its discoverers. The constants C_1 and C_2, originally thought to be universal constants, have been shown to vary rather slightly from polymer to polymer. A list of C_1 and C_2 for several of the most common polymers is presented in Table 3-2.

The form of equation (3-6) is straightforwardly predicted from rather simple theoretical considerations. First, however, we must realize that the shift factor a_T is not only meaningful in terms of

TABLE 3-2. WLF Parameters[2]

Polymer	C_1	C_2	T_g (°K)
Polyisobutylene	16.6	104	202
Natural rubber (Hevea)	16.7	53.6	200
Polyurethane elastomer	15.6	32.6	238
Polystyrene[6]	14.5	50.4	373
Poly(ethyl methacrylate)	17.6	65.5	335
"Universal constants"	17.4	51.6	

moduli. In Chapter 7, the equation

$$\eta(T) = \int_0^\infty E(T,t)\,dt \tag{3-7}$$

will be developed. $\eta(T)$ is the tensile viscosity of a polymer at the temperature T. Anticipating the results of Chapter 7, we use equation (3-7) in the argument which follows (Problem 7-8).

Consider two master curves at the different temperatures, T_g, which we will call *the* reference temperature, and T, any other temperature. We can write

$$E(T_g,t_{\text{ref}}) = E(T,t) \tag{3-8}$$

where we have constrained that every modulus value $E(T_g,t_{\text{ref}})$ has a corresponding modulus value of equal magnitude $E(T,t)$ on the master curve at temperature T. In writing this equation, we have implied a relationship between t_{ref} and t which, in the context of the above discussion, is realized to be

$$t_{\text{ref}} = \frac{t}{a_T} \tag{3-9}$$

where a_T is a function of T_g and T only. This is just another statement of time-temperature correspondence. (We are neglecting the vertical shifts for simplicity; their inclusion would substantially add to the complexity of this discussion but would not substantially affect the outcome.)

From equation (3-9) it is clear that

$$dt = a_T\,dt_{\text{ref}} \tag{3-10}$$

since a_T is not a function of time. Equation (3-7) becomes

$$\eta(T) = \int_0^\infty E(T,t)\,dt = \int_0^\infty E(T_g,t_{\text{ref}})\,dt \qquad (3\text{-}11)$$

where the last expression results from substitution of equation (3-8) into equation (3-7). In spite of the similarity of notation, t_{ref} and t are different variables as is obvious from equation (3-9). Elimination of dt via equation (3-10) yields

$$\eta(T) = \int_0^\infty E(T_g,t_{\text{ref}})a_T\,dt_{\text{ref}} \qquad (3\text{-}12)$$

where the limits of the integral are unchanged. Under this integration, however, a_T is a constant which can be factored out leaving an integral of the form of equation (3-7); thus

$$\eta(T) = a_T\eta(T_g)$$
$$a_T = \frac{\eta(T)}{\eta(T_g)} = \frac{\eta(T)}{\eta(T_{\text{ref}})} \qquad (3\text{-}13)$$

where the latter expression results from the fact that we arbitrarily chose T_g as the reference temperature.

We may now return to the theoretical rationalization of the form of the WLF equation. The starting point is the semiempirical Doolittle equation for the viscosity of a liquid[8]

$$\ln \eta = \ln A + B\left(\frac{V - V_f}{V_f}\right) \qquad (3\text{-}14)$$

which gives an expression for the viscosity of a system in terms of two constants A and B. We will assume this viscosity is the tensile viscosity. V is the total volume of the system while V_f is the free volume available to the system (a qualitative rather than quantitative view of free volume is sufficient for this discussion). The interpretation of equation (3-14) is that viscosity is intimately connected with mobility which, in turn, is closely related to free volume. As the free volume increases, the viscosity rapidly decreases. This equation has been found to express the viscosity dependence of simple liquids to a high degree of accuracy. Rearrangement of equation (3-14) gives

$$\ln \eta = \ln A + B\left(\frac{1}{f} - 1\right) \tag{3-15}$$

where f is the fractional free volume V_f/V. It is now assumed that above the glass transition temperature, the fractional free volume increases linearly, that is

$$f = f_g + \alpha_f(T - T_g) \tag{3-16}$$

where f is the fractional free volume at T, any temperature above T_g, f_g is the fractional free volume at T_g, and α_f is the thermal coefficient of expansion of the fractional free volume above T_g. In terms of equation (3-16) the Doolittle equation becomes

$$\ln \eta(T) = \ln A + B\left(\frac{1}{f_g + \alpha_f(T - T_g)} - 1\right) \quad \text{at } T > T_g$$

$$\ln \eta(T_g) = \ln A + B\left(\frac{1}{f_g} - 1\right) \quad \text{at } T_g \tag{3-17}$$

Subtraction yields

$$\ln \frac{\eta(T)}{\eta(T_g)} = B\left(\frac{1}{f_g + \alpha_f(T - T_g)} - \frac{1}{f_g}\right) \tag{3-18}$$

which simplifies to

$$\log \frac{\eta(T)}{\eta(T_g)} = \log a_T = -\frac{B}{2.303 f_g}\left(\frac{T - T_g}{(f_g/\alpha_f) + T - T_g}\right) \tag{3-19}$$

a form identical to the WLF equation where C_1 is identified with $(B/2.303 f_g)$ and C_2 with f_g/α_f. Thus the form of the WLF equation results from the accurate empirical Doolittle equation and the assumption of a linear expansion of free volume above T_g.

The complete viscoelastic response of any polymer under any experimental conditions may be obtained from a knowledge of any two of the following three functions: the master curve at any temperature, the modulus-temperature curve at any time and the shift factors relative to some reference temperature. For example, suppose we are given the constants C_1 and C_2 for a polymer whose master curve is known. For simplicity, we can assume that the master curve is at the same reference temperature as that in the WLF equation, perhaps T_g. Suppose it is desired to calculate the 10-second modulus versus temperature curve for this polymer. The 10-second modulus at T_g is read directly from the master curve. Now however, the master curve can be shifted to exhibit the behavior of the polymer at some other temperature. The amount of shift on a log scale, log a_T, is given by equation (3-6) where C_1 and C_2 used are those given. Applying this horizontal shift, with the slight additional vertical correction, if significant, allows one to "predict" the 10-second modulus, at this new temperature from the shifted curve. This procedure is repeated until the entire modulus-time curve is generated (Figure 3-7).

Having thus generated the manifold of curves shown in Figure 3-7, it is possible to view the dependence of the modulus-temperature plots on the arbitrary choice of time. This is done merely by picking

Figure 3-7. Plot of a Stress Relaxation Master Curve at Various Temperatures. Reprinted from A. V. Tobolsky and H. F. Mark, eds., *Polymer Science and Materials*, page 223 by permission of John Wiley & Sons, Inc.

Figure 3-8. Plot of Modulus Versus Temperature Curves for Various Values of the Constant Time. Reprinted from *Properties and Structure of Polymers* by A. V. Tobolsky, page 74 with the permission of John Wiley & Sons, Inc.

points off intersections of the master curves with vertical lines drawn from the point of interest on the time scale. The results of such treatment of the data[3] is shown in Figure 3-8. It is apparent that the longer the constant time of measurement, the sharper the resulting curves. In fact, if it is assumed that the ideas embodied in the WLF equation are applicable at temperatures considerably below T_g, it can be shown that an experiment of the type depicted in Figure 3-8 carried out infinitely slowly would result in a true second order thermodynamic transition, i.e., a discontinuity in the modulus, at a temperature about 50°C below T_g.

E. Molecular Interpretation of Viscoelastic Response

Finally we may briefly discuss the molecular interpretation of the behavior exhibited by polymers which are held at constant temperature and studied as a function of time.

In the very short time ranges, the molecules, not having sufficient time to reorient substantially, react to a stress probably by distorting intermolecular distances. These distortions, being of a rather high energy, result in a high modulus. At longer times, however, reorientation of chain segments is possible and occurs rather extensively. The chains reorient in order to relieve local strains imposed by the stress. Thus, severely pinched interactions will relax to lower energy conformations resulting in a modulus decrease. This is the glass transition process. At still longer times, the chains are, on the average, stretched. The average "end-to-end" length (Chapter 4) of the entangled chains has increased. This elongation is maintained due to long-range interactions which prevent long-range motions. Such a network of interacting flexible chains is a rubber. In the longest time range of interest, the chains can move by one another resulting in complete relaxation and a modulus indicative of a liquid. Now the end-to-end distance returns to its equilibrium value.

In this discussion, it is impossible to describe quantitatively the time ranges which give rise to each type of behavior since the temperature variable causes all of these ranges to be relative.

According to these ideas, a plastic would have the modulus of a rubber on a time scale of perhaps thousands of years while a rubber might behave like a plastic on a nanosecond time scale.

Direct evidence of this type of behavior is provided by the change in shape of the glass windows in some of the oldest cathedrals in Europe. On the time scale of many centuries, the glass is a rather viscous liquid; the bottoms of these windows are much thicker than the tops due to the response of this viscous liquid to the force of gravity. On the time scale of days, however, the material is hard, having withstood occasional high forces due to wind or pounding rain without appreciable response.

Problems

1. An apparent activation energy for viscoelastic relaxation may be obtained as the slope (multiplied by R) of a plot of $\ln a_T$ as a function of $1/T$. Such a plot yields a curved line and thus a temperature-dependent activation energy.[5]

(a) Obtain an analytical expression for the activation energy from the WLF equation. Using the "universal" constants C_1 and C_2, what is the value of the activation energy at T_g if (1) $T_g = 200°K$. (2) $T_g = 400°K$.

(b) Show that the energy becomes independent of temperature for $T \gg T_g$ and approaches the value of 4.1 kcal for all materials. Would you expect the activation energy for real polymers to obey this prediction? Why or why not?

2. The WLF equation may be written using any convenient temperature as reference temperature. The form of the equation remains he same but the values of the constants C_1 and C_2 change. Using he "universal" values for C_1 and C_2 at T_g, calculate C_1 and C_2 for a eference temperature of $T_g + 50°C$.

3. Tabulated below are the results of stress relaxation measurements made on bis-phenol-A-polycarbonate.[9] Construct the master curve and calculate C_1 and C_2, the WLF equation parameters, for a reference temperature of 150.8°C.

Stress Relaxation—Polycarbonate $M_w = 40,000$
(After Mercier, J. P. et al. J. Appl. Poly. Sci., **9**, 447 (1965))

T (°C)	$\log t$	$\log E(t)$	T (°C)	$\log t$	$\log E(t)$
130	2.00	10.25		4.00	9.66
	2.25	10.25		4.25	9.50
	2.50	10.25			
	2.75	10.25	142	2.00	10.13
	3.00	10.22		2.25	10.09
	3.25	10.19		2.50	10.03
	3.50	10.16		2.75	9.97
	3.75	10.13		3.00	9.88
	4.00	10.06		3.25	9.67
	4.25	10.03		3.50	9.63
				3.75	9.47
141	2.00	10.19		4.00	9.22
	2.25	10.19			
	2.50	10.16	144.9	2.25	9.66
	2.75	10.13		2.50	9.50
	3.00	10.09		2.75	9.28
	3.25	10.03		3.00	9.03
	3.50	9.94		3.25	8.88
	3.75	9.82		3.50	8.66

(continued)

Stress Relaxation—Polycarbonate $M_w = 40,000$ (*continued*)
(After Mercier, J. P. *et al. J. Appl. Poly. Sci.*, **9**, 447 (1965))

T (°C)	$\log t$	$\log E(t)$	T (°C)	$\log t$	$\log E(t)$
144.6	2.75	9.03		4.50	6.59
	3.00	8.72		4.75	6.38
	3.25	8.50		5.00	6.13
	3.50	8.25			
	3.75	8.09	159	2.25	7.34
	4.00	7.94		2.50	7.25
				2.75	7.16
146	2.50	8.72		3.00	7.03
	2.75	8.47		3.25	6.94
	3.00	8.25		3.50	6.78
	3.25	8.06		3.75	6.63
	3.50	7.94		4.00	6.41
	3.75	7.82		4.25	6.13
	4.00	7.67		4.50	5.85
150.8	2.00	7.97		4.75	5.53
	2.25	7.85	161.5	2.25	7.06
	2.50	7.78		2.50	6.97
	2.75	7.72		2.75	6.85
	3.00	7.69		3.00	6.64
	3.25	7.64		3.25	6.50
	3.50	7.59		3.50	6.31
	3.75	7.55		3.75	6.06
	4.00	7.50		4.00	5.78
	4.25	7.44		4.25	5.47
	4.50	7.34			
	4.75	7.25	167	2.25	6.72
156	2.00	7.59		2.50	6.50
	2.25	7.55		2.75	6.25
	2.50	7.50		3.00	5.97
	2.75	7.41		3.25	5.63
	3.00	7.34		3.50	5.22
	3.25	7.25	171	2.25	6.00
	3.50	7.19		2.50	5.72
	3.75	7.06		2.75	5.41
	4.00	6.94		3.00	5.03
	4.25	6.78		3.25	4.63

References

1. A. V. Tobolsky, *Properties and Structure of Polymers*, Wiley, New York, 1960.

2. J. D. Ferry, *Viscoelastic Properties of Polymers*, Wiley, New York, 1970.

3. H. Fujita, *J. Colloid Sci.*, **12,** 204 (1957).

4. A. V. Tobolsky and J. R. McLoughlin, *J. Polymer Sci.*, **8,** 543 (1952).

5. E. Catsiff and A. V. Tobolsky, *J. Colloid Sci.*, **10,** 375 (1955).

6. D. J. Plazek, *J. Phys. Chem.*, **69,** 3480 (1965).

7. M. L. Williams, R. F. Landel, and J. D. Ferry, *J. Amer. Chem. Soc.*, **77,** 3701 (1955).

8. A. K. Doolittle and D. B. Doolittle, *J. Appl. Phys.*, **31,** 1164 (1959).

9. J. P. Mercier *et al.*, *J. Appl. Polymer Sci.*, **9,** 447 (1965).

4 Transitions and Relaxation in Amorphous Polymers

Amorphous polymers, as we have seen in Chapter 3, undergo a transition from glassy behavior to rubbery behavior. This transition is quite sharp, usually occurring over a temperature range of a few degrees. The decrease in modulus of about 3 orders of magnitude which accompanies this transition makes it clear that it is the single most important parameter characterizing the mechanical behavior of amorphous polymers.

In this chapter, the glass transition phenomenon will be treated in detail including phenomenological aspects, molecular theories, and the effect of molecular structure. In addition, relaxation occurring in the glassy state below the glass transition temperature, (T_g), will be discussed.

A. Phenomenology of the Glass Transition

The classic method for the experimental determination of the glass transition temperature is dilatometry. Thus, as briefly mentioned in Chapter 1, the temperature dependence of the specific volume is determined by a suitable technique and the temperature at the change in slope is taken as T_g. Such a plot is indicated schematically in Figure 4-1 where it is shown that the T_g is actually determined as the intersection of the straight line portions of the curve above and below T_g. The slope is, of course, related to the cubical coefficient of thermal expansion, α. Figure 4-2 shows that a discontinuity in α is observed at T_g.

As is apparent from Figures 4-1 and 4-2, the glass transition is a rate-dependent phenomenon. Thus a rapid heating or cooling rate leads to a higher T_g than does a slow rate.

Dilatometric methods are simple in principle but complex in practice. The displacement method is perhaps most commonly used. The thermal expansion coefficient is determined by measuring the

Figure 4-1. Schematic Plot of the Temperature Dependence of the Specific Volume or Enthalpy of an Amorphous Polymer. Effects of Different Cooling Rates Are Included.

Figure 4-2. Schematic Plot of the Temperature Dependence of the Thermal Expansion Coefficient or Heat Capacity for an Amorphous Polymer. Effects of Different Cooling Rates Are Included.

amount of confining fluid displaced by the polymer. The success of the method depends on the selection of a confining fluid which is not absorbed by the polymer and undergoes no phase transitions in the temperature range of interest. Mercury is often used although it possesses obvious temperature limitations.

In addition to dilatometry, calorimetry has been extensively employed, a discontinuity in heat capacity being observed at T_g (Figure 4-2). The recent developments of differential thermal analysis and differential scanning calorimetry[1] have enhanced the importance of thermodynamic measurements in the determination of T_g.

It has been repeatedly emphasized throughout this book that the glass transition in amorphous polymers is accompanied by profound changes in their viscoelastic response. Thus the stress relaxation modulus commonly decreases by about three orders of magnitude in the vicinity of the T_g determined by calorimetry or dilatometry, and the creep compliance increases by about three orders of magnitude. In addition, under dynamic experimental conditions, the storage moduli and compliances behave in a similar manner to the corresponding static quantities. The loss moduli and compliances, on the other hand, exhibit maxima in the glass transition region, as does tan δ. Figure 4-3 summarizes these results for a styrene-butadiene copolymer. The data in Figure 4-3 were collected at a constant frequency of 1 cps over the temperature range indicated. (The logarithmic decrement, Δ, is plotted in Figure 4-3 rather than tan δ, but this is merely related to tan δ through the relationship $\Delta \approx \pi \tan \delta$.) It can be seen from Figure 4-3 that the maxima in tan δ and G'' occur at different temperatures, the G'' maximum being lower. Frequently, the tan δ or G'' peak temperatures are

Figure 4-3. Temperature Dependencies of G', G'', and the Logarithmic Decrement, Δ, for a Styrene Butadiene Copolymer. After L. E. Nielsen, *Mechanical Properties of Polymers*, Reinhold, New York, 1962, page 128.

taken as the T_g but it is clear that they will not be identical and that in general they will both yield higher values than those obtained by dilatometric or thermodynamic methods, owing to differences in the time scale. Figure 4-4 gives the temperature dependence of E' and tan δ for polyvinyl chloride at various frequencies. The shift of the tan δ peak maximum to higher temperatures with increasing frequency occurs in accordance with the WLF equation and is again an illustration of the rate dependence of the glass transition phenomenon.

One of the oldest methods for the determination of T_g involves the temperature dependence of viscosity. Figure 4-5 is a plot of the variation of viscosity with temperature for polymethyl methacrylate. Various authors have argued that the glass transition temperature

Figure 4-4. Temperature Dependencies of E' and tan δ for Polyvinyl Chloride at Various Frequencies. After G. W. Becker *Kolloid Z*, **140**, 1 (1955). Reprinted with the permission of Dr. Dietrich Steinkopff Verlag.

Figure 4-5. Temperature Dependence of the Viscosity for Polymethyl Methacrylate with $M = 63,000$. Reprinted from *Physical Properties of Polymers* by F. Bueche, page 101 with the permission of John Wiley & Sons, Inc.

represents an isoviscous state with viscosity 10^{13} poises. Such a high viscosity is not easily measurable by normal flow techniques and, in fact, its determination is usually accomplished from creep or stress relaxation measurements by the techniques of Chapter 7.

Many profound physical changes occur at T_g in addition to those already mentioned. The refractive index undergoes an abrupt change at T_g. In the case of polar polymers the dielectric loss tangent passes through a maximum in the vicinity of T_g as does the imaginary part of the complex dielectric constant. Because of the increased mobility at T_g, the line width in a nuclear magnetic resonance experiment undergoes an abrupt narrowing at this temperature. All of these effects, and many others, have been used as measures of T_g, and they all arise from the fundamental changes in molecular dynamics which occur in the vicinity of T_g. It should be apparent from the discussion that each experimental method will yield a different value for T_g and that even the same method will yield different results depending on the time scale. Thus, in order to compare T_g values, both the method of measurement and the rate of measurement should be specified.

B. Theories of the Glass Transition

1. FREE VOLUME THEORY

In Chapter 3 the WLF equation was derived on the basis of free volume concepts. In particular, we may write

$$\log a_T = \frac{B}{2.303}\left(\frac{1}{f_T} - \frac{1}{f_{T_0}}\right) \tag{4-1}$$

where f_T and f_{T_0} are the fractional free volumes at temperatures T and T_0, respectively, and B, a parameter in the Doolittle equation, is experimentally found to be close to unity.

Unfortunately, the concept of free volume is difficult to define in a precise manner. In an approximate way we can represent the segments of a polymer chain by rigid bodies and the free volume as the holes present between these segments as a result of packing requirements. Presumably the free volume reaches a constant value at T_g which is too small to allow the large-scale conformational rearrangements of the chain backbones associated with T_g to occur. Above T_g, on the other hand, free volume increases and becomes sufficiently large to allow such motions to occur. These ideas are embodied in equation (3-16) which we requote here.

$$f = f_g + \alpha_f[T - T_g] \qquad T \geq T_g \tag{3-16}$$
$$f = f_g \qquad\qquad\quad T < T_g$$

where the fractional free volume reaches a constant value, f_g, at T_g and increases linearly above T_g with the coefficient of expansion α_f.

Following Chapter 3, substitution of equation (3-16) into equation (4-1) with T_g as the reference temperature yields the WLF equation where the constants C_1 and C_2 are given by

$$C_1 = \frac{B}{2.303 f_g} \qquad C_2 = \frac{f_g}{\alpha_f} \qquad (4\text{-}2)$$

Knowledge of the numerical values of C_1 and C_2 thus leads to the parameters f_g and α_f through equation (4-2), since B is taken as unity. The constants C_1 and C_2 were originally taken to be universal for all amorphous polymers with $C_1 = 17.44$ and $C_2 = 51.6$. It was later found that C_1 values were indeed approximately constant for all systems but that C_2 varied quite widely. This is illustrated by Table 3-2. Equation (4-2) indicates that this result means that f_g is approximately constant at a value of 0.025, but that α_f varies from one amorphous polymer to another. It is clear that the WLF equation predicts that T_g represents an iso-free volume state. While this concept is not strictly true it is nevertheless of wide utility as we shall see in the following discussion.

It is possible to identify α_f with $\Delta\alpha = \alpha_r - \alpha_g$, that is, the difference between the expansion coefficients above and below T_g. If this is done, good agreement is obtained for many polymers between experimentally determined values of $\Delta\alpha$ and α_f obtained from C_2 values through equation (4-2). Inasmuch as there is no *a priori* method of defining α_f, the above agreement is perhaps the best justification for the identification of α_f with $\Delta\alpha$.

The free volume approach embodied in equation (3-16) may be easily generalized to include the effect of pressure. Intuitively, pressure may be thought of as having the effect of "squeezing out" free volume and thus raising the glass transition temperature and such has actually been found to be the case. The magnitude of the effect is predictable as follows. Assume the polymer is taken from state 1 characterized by P_1 and T_{g_1}, to state 2 characterized by T_{g_2} and P_2. We may write in general

$$f_2 = f_1 + \alpha_f[T - T_0] - \beta_f[P - P_0] \qquad (4\text{-}3)$$

In equation (4-3) β_f refers to the isothermal compressibility of free volume, assumed independent of pressure over the range of interest, and T_0 and P_0 are the reference pressure and temperature. If the polymer remains at T_g in going from state 1 to state 2, f_2 must also remain equal to f_1 since T_g is an iso-free volume state. Thus, under these conditions

$$\alpha_f[T - T_0] = \beta_f[P - P_0] \tag{4-4}$$

For small changes, equation (4-4) becomes

$$\left[\frac{dT}{dP}\right]_{T=T_g} = \frac{\beta_f}{\alpha_f}$$

or

$$\frac{dT_g}{dP} = \frac{\beta_f}{\alpha_f} \tag{4-5}$$

Let $\beta_f = \Delta\beta$, the compressibility above and below T_g, and $\alpha_f = \Delta\alpha$. Then

$$\frac{dT_g}{dP} = \frac{\Delta\beta}{\Delta\alpha} \tag{4-6}$$

Although precise pressure measurements are difficult and not very much data have been reported, amorphous polymers at least approximately obey equation (4-6). (See Table 4-1.)

The free volume theory finds ready application in predicting the effect on T_g of diluents and molecular weight among others. As an example, we shall derive an expression for the dependence of T_g on plasticizer concentration. The total fractional free volume of a

TABLE 4-1

Polymer	$TV \Delta\alpha/\Delta C_p$	dT_g/dP	$\Delta\beta/\Delta\alpha$
Polyvinyl acetate	0.025[a]	0.02[a]	0.05[a]
Polystyrene	—	0.036[b]	0.10[b]
Natural rubber	0.020[a]	0.024[c]	0.024[c]
Polymethyl methacrylate	—	0.023[d]	0.065[e]

[a] J. M. O'Reilly, *J. Polymer Sci.*, **5**, 429 (1962).
[b] S. Matsuoka and B. Maxwell, *J. Polymer Sci.*, **32**, 131 (1958).
[c] J. E. McKinney, H. V. Belcher, and R. S. Marvin, *Trans. Soc. Rheo.*, **4**, 347 (1960).
[d] N. Shishkin, *Soviet Phys.—Solid State*, **2**, 322 (1960).
[e] G. Allen, D. Sims, and G. J. Wilson, *Polymer*, **2**, 315 (1961).

polymer-diluent system may be written as

$$f = 0.025 + \alpha_p(T - T_{g_p})V_p + \alpha_d(T - T_{g_d})V_d \tag{4-7}$$

In equation (4-7) the subscripts p and d refer to polymer and diluent, respectively, and V is the volume fraction. At the T_g of the polymer-diluent mixture, f becomes 0.025. Using this fact and substituting $1 - V_p$ for V_d, we have

$$T_g = \frac{\alpha_p V_p T_{g_p} + \alpha_d(1 - V_p)T_{g_d}}{\alpha_p V_p + \alpha_d(1 - V_p)} \tag{4-8}$$

Figure 4-6. Dependence of T_g on V_p for Mixtures of Polymethyl Methacrylate with Diethyl Phthalate. Comparison of Experimental Results with Equation (7-8). Parameters Used Are $\alpha_d = 10 \times 10^{-4}$ and $T_{gd} = -65°C$. After M. Gordon and J. S. Taylor, *J. Appl. Chem.*, **2**, 493 (1952) with the permission of the Society of Chemical Industry.

The dependence of T_g on V_p as predicted from equation (4-8) is plotted. in Figure 4-6 for the system polymethyl methacrylate-diethyl phthalate, together with some experimental results. The calculated curve is seen to agree well with the experimental data.

2. THERMODYNAMIC THEORY

Equilibrium phase transitions are well treated using the classical thermodynamic approach. In particular, the phenomena of melting and vaporization are characterized by the equality of the Gibbs Free Energy of the two phases at the transition temperature.

$$F_1 = F_2 \tag{4-9}$$

Also

$$dF_1 = dF_2 \tag{4-10}$$

However, the volumes and entropies of the two phases are not equal.

$$\begin{aligned} S_1 &\neq S_2 \\ V_1 &\neq V_2 \end{aligned} \tag{4-11}$$

Ehrenfest[2] refers to the type of phase transition described above as a first order transition because there are discontinuities in the first partial derivatives of the Gibbs Free Energy at the transition point. For example,

$$dF = -S\,dT + V\,dP \tag{4-12}$$

and we have

$$\begin{aligned} S &= -\left(\frac{\partial F}{\partial T}\right)_P \\ V &= \left(\frac{\partial F}{\partial P}\right)_T \end{aligned} \tag{4-13}$$

This notion is easily generalized to higher order transitions. Thus a second order transition is described as one in which the second partial derivatives of the Gibbs Free Energy show discontinuities at the transition point, a third order transition involves discontinuities in the third partial derivatives, etc.

In particular, for a second order transition,

$$-\left(\frac{\partial^2 F}{\partial T^2}\right)_P = \left(\frac{\partial S}{\partial T}\right)_P = \frac{C_p}{T}$$

$$\left(\frac{\partial^2 F}{\partial P^2}\right)_T = \left(\frac{\partial V}{\partial P}\right)_T = -\beta V \qquad (4\text{-}14)$$

$$\left(\frac{\partial}{\partial T}\left(\frac{\partial F}{\partial P}\right)_T\right)_P = \left(\frac{\partial V}{\partial T}\right)_P = \alpha V$$

So, at the second order transition temperature, the following discontinuities are observed

$$\Delta C_p = C_{p_2} - C_{p_1}$$

$$\Delta\beta = \beta_2 - \beta_1 \qquad (4\text{-}15)$$

$$\Delta\alpha = \alpha_2 - \alpha_1$$

It is just these discontinuities that occur in the case of the glass transition, and, because of this, T_g is often referred to as the second order transition point. It is clear, however, that the observed T_g is a rate-dependent phenomenon. The thermodynamic analysis given above refers only to an equilibrium process and thus can not be directly applied to the experimental T_g. The kinetic nature of the observed T_g does not, however, preclude the existence of a true second order transition temperature. Thus in a dilatometric experiment, when the polymer is cooled from the rubbery or liquid states, volume contractions take place involving conformational rearrangements. At temperatures far above T_g thermal equilibrium is maintained during the cooling process, but as the temperature is lowered a point is eventually reached at which the rate of volume contraction becomes comparable to the rate of cooling. Below this temperature, volume relaxation and hence conformational rearrangements are not possible during the time scale of the experiment and the discontinuities ΔC_p, $\Delta\beta$, and $\Delta\alpha$ are observed. Such an analysis indicates that an infinitely slow cooling rate would be required to observe the true thermodynamic second order transition temperature, if it exists at all.

Kauzmann[3] noted that if the entropies of simple glass forming liquids were extrapolated to low temperatures, they would go to zero long before the absolute zero of temperature was reached. Such a result, requiring negative entropies, is clearly meaningless in a physical sense. Kauzmann resolved this paradox by suggesting that the glassy state is not an equilibrium state and that the glasses would pass over into crystalline solids at equilibrium before the temperature of zero extrapolated entropy could be achieved. Such an explanation denies the existence of a second order thermodynamic transition. However, many polymeric substances have never been obtained in a crystalline state and it is difficult to imagine the possibility of crystallizing a material such as atactic polystyrene, for example. It was subsequently suggested by Gibbs and DiMarzio[4] that the conformational entropy at *equilibrium* would indeed attain a value of zero at a finite temperature, defined as T_2, undergoing no further change between T_2 and the absolute zero. This idea is schematically illustrated in Figure 4-7 in which the dotted line represents Kauzmann's extrapolation. The T_2 of Gibbs and DiMarzio is a true second order transition temperature. In this view the observed T_g is the kinetic reflection of the underlying thermodynamic phenomenon which would occur at T_2.

Figure 4-7. Schematic Plot of Conformational Entropy Versus Temperature for a Glass-Forming Substance. The Temperature at which the Entropy Attains a Value of 0 Is the T_2 of Gibbs and DiMarzio. The Dotted Line Is the Original Extrapolation of Kauzmann. After J. H. Gibbs in *Modern Aspects of the Vitreous State*, Vol. 1. Reprinted with the permission of Butterworth and Co.

At high temperatures, Gibbs and DiMarzio suggest that a very great number of conformations are accessible to each chain. They assume that for each chain segment there is one definite lowest energy conformation, so that at equilibrium at low temperatures there is just *one* conformation available to the entire chain. Thus, as the chain is cooled from high to low temperatures, fewer and fewer high energy conformations become accessible until at T_2 only the one lowest energy state is allowed.

In reality however, there are so many states accessible at high temperatures that the reorientation from one state to any other particular state may take quite some time. In fact, as we cool the sample and intrinsically slow down molecular movement, it should take longer and longer to get to any particular conformation. To insure getting all the chains into the lowest energy conformation, the experiment would have to be carried out infinitely slowly.

The temperature at which the second order transition would occur according to these ideas may be calculated using the WLF equation (3-6). To shift an experiment from a finite time scale to an infinite time scale would take a value of log a_T approaching infinity. Clearly this will be the case if the denominator of the right hand side of equation (3-6) goes to zero while the numerator remains finite.

$$C_2 + T_2 - T_g = 0 \qquad (4\text{-}16)$$

Since C_2 is relatively constant for most polymers, one derives the relation that

$$T_2 = T_g - C_2 \approx T_g - 52 \qquad (4\text{-}17)$$

Thus the second order phase change woud be observed about 50°C below T_g for an experiment carried out infinitely slowly.

The Gibbs-DiMarzio model is characterized by two parameters. These are the hole formation energy, u_0, and the flex energy, ε. u_0 is a reflection of intermolecular energy contributions and ε is a reflection of intramolecular energy contributions. Specifically, for hydrocarbon chains such as polymethylene, u_0 is the energy of interaction, or the "van der Waals bond" energy between a pair of chemically nonbonded but nearest-neighboring segments in the lattice. In order to define ε, we note that if Z is the number of primary valences of each backbone chain atom, there are $(Z - 1)$ possible orientations of a bond i with respect to the coordinate system formed by the bonds $(i-1)$ and $(i-2)$ of the same molecule. In the case of a hydrocarbon chain, $Z = 4$. It is assumed that an energy ε_1 is associated with one of the three possible orientations and an energy ε_2 is associated with the other two, with $\varepsilon_2 > \varepsilon_1$. Then ε is defined as the difference between ε_2 and ε_1. It should be emphasized that ε, a thermodynamic quantity, has nothing to say about the magnitude of potential energy barriers to rotation about chain backbone bonds. Rather ε refers only to the energy difference between low lying rotational isomeric states.

This theory predicts that V_0, the unoccupied volume at T_2, is a function of the product $T_g \Delta \alpha$ alone. Simha and Boyer[5] have shown that this product is a constant for amorphous polymers and thus one can calculate that $V_0 = 0.025$ for such polymers at T_2. This fraction is, of course, identical to the value of f_g in the free volume theory. The agreement is merely fortuitous, however, inasmuch as f at T_2 is taken to be zero in the free volume approach. In addition, according to the theory, it is found that E_0 is directly proportional to T_2 and that ε/kT_2 is a constant for all polymers. Recasting this ratio in terms of T_g rather than T_2, it can be shown that ε/kT_g possesses the "universal" value of 2.26 for all amorphous polymers.

We shall demonstrate the utility of the Gibbs-DiMarzio theory by applying it to the calculation of the T_g of a random copolymer. To do this, we assume that the flex energy of a random copolymer composed of A and B monomer units is given by

$$\varepsilon = X_A \varepsilon_A + X_B \varepsilon_B \qquad (4\text{-}18)$$

where X_A and X_B are the mole fractions of monomer units A and B, respectively. Substituting the "universal" value of $\varepsilon = 2.26 kT_g$ in equation (4-18) immediately leads to

$$T_g = X_A T_{g_A} + X_B T_{g_B} \qquad (4\text{-}19)$$

which is identical in form to an empirical equation of Wood[6] found to have wide applicability to random copolymers.

The Gibbs-DiMarzio approach, as has been stated, rests on the existence of a true second order thermodynamic transition at a temperature T_2 below the observed T_g. The pressure dependence of

T_2 can thus be obtained by the methods of equilibrium thermo-dynamics. For a first order phase transition, the pressure dependence of the transition temperature is given by the Clapeyron equation

$$\frac{dT}{dP} = \frac{\Delta V}{\Delta S} \qquad (4\text{-}20)$$

Straightforward application of the Clapeyron equation to a second order transition is not possible because both ΔV and ΔS are 0, and thus dT/dP is indeterminate. However, we may invoke L'Hopital's rule and differentiate both numerator and denominator of the right hand side of equation (4-20) independently to obtain the limiting behavior. This may be done either with respect to temperature or pressure. Thus, recalling equation (4-14) and differentiating equation (4-20) with respect to temperature

$$\frac{dT_2}{dP} = \frac{\partial \Delta V/\partial T}{\partial \Delta S/\partial T} = \frac{V \, \Delta\alpha T_g}{\Delta C_p} \qquad (4\text{-}21)$$

Differentiating with respect to pressure there results

$$\frac{dT_2}{dP} = \frac{\partial \Delta V/\partial P}{\partial \Delta S/\partial P} = \frac{\Delta\beta}{\Delta\alpha} \qquad (4\text{-}22)$$

Thus the pressure dependence of T_2 and hence T_g is given by either equation (4-21) or equation (4-22). Equation (4-22) is, of course, identical to the free volume result, equation (4-6). A comparison of these quantities has been given in Table 4-1.

C. Structural Parameters Affecting the Glass Transition

We have seen that the free volume and thermodynamic theories serve to rationalize the glass transition phenomenon in a wide variety of polymeric systems. There are, of course, additional effects which cannot be well explained on the basis of either of these approaches. In this section we examine some structural parameters affecting T_g in a qualitative manner, pointing out exceptions to the theories quoted above.

TABLE 4-2. Glass Transition Temperatures for Selected Polymers

Organic Polymer	T_g (°C)
Polyacenaphthalene	264
Polyvinyl pyrrolidone	175
Poly-o-vinyl benzyl alcohol	160
Poly-p-vinyl benzyl alcohol	140
Polymethacrylonitrile	120
Polyacrylic acid	106
Polymethyl methacrylate	105
Polyvinyl formal	105
Polystyrene	100
Polyacrylonitrile	96
Polyvinyl chloride	87
Polyvinyl alcohol	85
Polyvinyl acetal	82
Polyvinyl proprional	72
Polyethylene terephthalate	69
Polyvinyl isobutyral	56
Polycaprolactam (nylon 6)	50
Polyhexamethylene adipamide (nylon 6,6)	50
Polyvinyl butyral	49
Polychlorotrifluorethylene	45
Ethyl cellulose	43
Polyhexamethylene sebacamide (nylon 6,10)	40
Polyvinyl acetate	29
Polyperfluoropropylene	11
Polymethyl acrylate	9
Polyvinylidene chloride	−17
Polyvinyl fluoride	−20
Poly-1-butene	−25
Polyvinylidene fluoride	−39
Poly-1-hexene	−50
Polychloroprene	−50
Polyvinyl-n-butyl ether	−52
Polytetramethylene sebacate	−57
Polybutylene oxide	−60
Polypropylene oxide	−60
Poly-1-octene	−65
Polyethylene adipate	−70
Polyisobutylene	−70
Natural rubber	−72
Polyisoprene	−73
Polybutadiene	−85
Polydimethyl siloxane	−123

TABLE 4-2. (*Continued*)

Inorganic Polymers	T_g (°C)
Silicon dioxide	1200–1700
Polycalcium phosphate	525
Polysodium phosphate	285
Boron trioxide	200–260
Arsenic trisulfide	195
Arsenic trioxide	160
Zinc chloride	100
Sulfur	75
Selenium	30
Polyphosphoric acid	−10

The details of the molecular structure of polymers profoundly influence the observed T_g's as illustrated by Table 4-2 where we may contrast the T_g of polydimethyl siloxane, −123°C, with that of polycalcium phosphate, +525°C. At least approximately, we may separate the observed effects into intermolecular and intramolecular parts. The latter refer to structural parameters affecting the stiffness of the chain backbone and we shall examine these first.

The internal mobility of a chain reflects the ease of rotation of backbone bonds about one another. This is determined by the barrier to internal rotation and by steric hindrance introduced by the presence of substituents on the backbone chain atoms. Thus, the low T_g of polydimethyl siloxane may be rationalized on the basis of a low rotational energy barrier. A similar explanation may be advanced for the low T_g of linear polyethylene (−120°C). [Some controversy still exists concerning the T_g of linear polyethylene. The value of −120°C is adopted here from the recent work of Stehling and Mandelkern[7].] The steric hindrance factor may be illustrated by considering the effect of the substitution of various groups for hydrogen atoms on a linear polyethylene chain. The placement of a methyl group on alternate carbon atoms results in polypropylene which has a T_g of −20°C. Polystyrene, which has a phenyl group on alternate carbon atoms has a T_g of +100°C, over 100° higher than that of unsubstituted polymethylene. This type of correlation must be applied with caution, however. Consider polyisobutylene, containing two methyl groups on alternate carbon atoms. In this case the T_g is

−80°C, some 60° lower than the monosubstituted case, polypropylene.

An increase in side group bulkiness generally serves to raise T_g. Ring substituted polystyrenes show this effect and, in the case of polyacenaphthalene, the mobility of the chain is so severely inhibited that the T_g occurs at 264°C.

Chain microstructure is clearly important in determining chain mobility. We have already discussed the case of random copolymers on the basis of the Gibbs-DiMarzio theory and have developed equation (4-19) which shows that the T_g of a random copolymer is intermediate between the T_g's of the corresponding homopolymers. Equation (4-19) is identical in form to the two-parameter empirical expression of Wood[6].

$$K_A W_A (T_g - T_{g_A}) + K_B W_B (T_g - T_{g_B}) = 0 \qquad (4\text{-}23)$$

In equation (4-23), W_A and W_B are the weight fractions of the comonomers A and B and K_A and K_B are constants characteristic of A and B, respectively.

Equation (4-23) may be rearranged to

$$T_g = [T_{g_A} + (kT_{g_B} - T_{g_A})W_B]/[1 - (1 - k)W_B] \qquad (4\text{-}24)$$

where $k = K_B/K_A$.

Equation (4-24) is known as the Gordon-Taylor[8] equation and has found wide application to random amorphous copolymers. Figure 4-8 shows the experimental results for a series of styrene-butadiene copolymers along with the corresponding T_g's calculated from equation (4-24) with $k = 0.28$. The above equations (4-23) and (4-24) are applicable only to random copolymers. Very different phenomena are observed in the case of block and graft copolymers. Frequently in these situations, two T_g's are observed as a consequence of the occurrence of microphase separation. This subject is beyond the scope of this book, however.

In addition to copolymer composition, geometrical and steric isomerism play important roles in the determination of chain stiffness and thus of T_g. Polydienes, for example, can exist as cis and trans geometrical isomers. In the case of poly-1,4-butadiene, the cis isomer has a T_g of −102°C while the stiffer trans isomer shows a T_g of

WT. FRACTION OF STYRENE.

Figure 4-8. Composition Dependence of T_g for a Series of Styrene-Butadiene Copolymers Compared to the Predicted Curve Calculated on the Basis of Equation (4-24) with $k = 0.28$. After F. Bueche *et. al.*, *J. Polymer Sci.*, **50,** 549 (1961). Reprinted with the permission of John Wiley & Sons, Inc.

$-48°C$. The effect is not nearly so marked in the case of polyisoprene where the cis isomer exhibits a T_g of $-73°C$ and the trans isomer a T_g of $-53°C$.

Vinyl polymers of the type (CH_2CHX) and vinylidene polymers of the type (CH_2CXY) are capable of existing in different stereoregular forms. These are generally referred to as isotactic, syndiotactic, and atactic forms and it is here assumed that the reader is familiar with this nomenclature and the specific structures associated with it. It has been shown that steric isomerism has no effect on the T_g of vinyl polymers but profoundly affects the T_g of vinylidene polymers.[5] In the latter case, the isotactic form invariably is associated with the lowest T_g values. This result has also been rationalized on the basis of the Gibbs-DiMarzio theory but again the details are beyond the scope of this book.

Turning to intermolecular effects we first note that the free volume approach is of great utility here, although by no means universally applicable. One example is the effect of low molecular weight diluents or plasticizers which have already been accounted for by the free volume theory through equation (4-8). The general idea is that plasticizers lower T_g by introducing free volume into the system, the final T_g being intermediate between that of the plasticizer and that of the polymer. The free volume theory cannot account for the plasticizing effect of water on polar polymers, however. Water exerts a dramatic plasticizing effect on nylons, greater than 30°C for the first weight per cent. However, the addition of water in low concentrations actually increases the density of the polymer water mixture and would thus be expected to raise the T_g on the basis of the free volume theory rather than lowering it as actually observed.

Figure 4-9. T_g Versus Reciprocal Molecular Weight for Polystyrene Fractions. The Solid Line Is a Plot of Equation (4-25) with $c = 1.7 + 10^5$. After T. G. Fox and P. J. Flory, *J. Appl. Phys.*, **21**, 581 (1950), with the permission of the American Institute of Physics.

The explanation is that water acts to break intermolecular hydrogen bonds present in the nylon, thus allowing greater chain mobility and also more efficient packing, the net result leading to a decrease in T_g.

The effect of molecular weight has been alluded to in Chapter 3. Specifically, free volume around chain ends is taken to be greater than that around chain middles because of imperfect packing at the chain ends. It is expected that T_g will become independent of molecular weight at high molecular weights due to the negligible concentration of chain ends in high molecular weight polymers. It follows directly from the free volume theory (see Problem 6-1) that

$$T_g = T_g^\infty - c/\overline{M}_n \qquad (4\text{-}25)$$

which is known as the Fox-Flory equation.

Figure 4-9 illustrates equation (4-25) for the case of polystyrene and it is seen that the data are well fitted for $c = 1.7 \times 10^5$. The c values vary from polymer to polymer but are generally of the same order of magnitude as the polystyrene case.

D. Relaxations in the Glassy State

In addition to the glass transition, amorphous polymers usually exhibit at least one so-called secondary relaxation region. Secondary relaxations are manifestations of motions within the polymer in the glassy state. Since large-scale motions such as those accompanying the glass transition are impossible in the glassy state, the secondary relaxations must arise from localized motions. These are conveniently classified into two types: main chain motions and side group motions. In this section we shall describe methods of detecting secondary relaxations and give examples of secondary relaxations belonging to both side group and main chain motions.

Until very recently, no secondary relaxation had been observed by calorimetry or dilatometry. It has now been reported that the side chain relaxation in polymethyl methacrylate has been detected dilatometrically.[10] The necessary sensitivity to detect these relaxations by the dilatometric method is very difficult to achieve. These remarks also apply to the static viscoelastic measurements such as creep and stress relaxation. There is only a single report of the detection of a secondary relaxation by creep, again the side group relaxation in polymethyl methacrylate[11] and no reports of any detected by stress relaxation.

Dynamic methods, on the other hand, easily detect glassy state relaxations and have been extensively applied to their study. These include dynamic mechanical methods, dielectric relaxation, and nuclear magnetic resonance. Since we are primarily concerned with viscoelastic response in this book, we shall confine the discussion to the dynamic mechanical technique. As should be apparent, it is possible to detect secondary relaxations by performing measurements

Figure 4-10.   Temperature Dependence of tan δ for Polymethyl Methacrylate Showing the α and β Relaxations. After L. E. Nielsen, *Mechanical Properties of Polymers*, Reinhold, New York, 1962, page 178.

over a frequency range at constant temperature or over a temperature range at constant frequency. The latter technique is usually employed, largely because extended frequency ranges are difficult to achieve experimentally and many secondary relaxations of interest occur at very high frequencies at room temperature.

As an example may be cited the case of polymethyl methacrylate. Figure 4-10 shows a plot of the logarithmic decrement versus temperature for this polymer at a constant frequency of 1 cps in the temperature range $-50°$ to $+160°C$. Two relaxation peaks are discernible, a relaxation corresponding to the glass transition at $110°C$ and a secondary relaxation at $50°C$. It becomes necessary to adopt a labeling scheme for the various relaxations observed. Several schemes are currently in use and we shall choose one of the most popular, that of labeling the peaks α, β, γ, etc. in order of decreasing temperature. Thus, in a linear amorphous polymer, the α relaxations corresponds to the glass transition, the β relaxation to the highest temperature glassy state relaxation, and so forth. Polymethyl methacrylate exhibits only one secondary relaxation in the temperature range of Figure 4-10 and this is the β relaxation at $50°C$.

Much of the discussion of the motions underlying the glass transition has been presented without reference to details of molecular structure. Because of the relatively long-range motions characterizing the glass transition, it is possible to construct general theories applicable to all amorphous polymers without regard to fine structural details. Such is not the case with the localized secondary relaxations. We must in general consider structural features concerning only a few main chain or side group atoms. Many different mechanisms have been proposed for the large variety of secondary relaxations observed, and a detailed consideration is beyond the scope of this treatment. A very complete account exists in the work of McCrum, Read, and Williams[12] which should be consulted for more details. We shall present only two mechanisms here.

First is the crankshaft mechanism of Schatzki.[13] It has been observed for many polymers containing linear $(CH_2)_n$ sequences with $n = 4$ or greater, that a secondary relaxation occurs at about $-120°C$ at 1 cps. This seems to be true regardless of whether the CH_2 sequences occur in the main chain or in the side groups. Thus

Figure 4-11.   The Crankshaft Mechanism According to Schatzki.

both polyethylene and poly-n-butyl methacrylate exhibit this relaxation. The mechanism proposed by Schatzki is shown in Figure 4-11.

The motion responsible for the relaxation is a rotation about the two colinear bonds 1 and 7 such that the carbon atoms between bonds 1 and 7 move in the manner of a crankshaft. The colinearity of the two terminal bonds is achievable if there are four intervening carbon atoms on the assumption of tetrahedral valence angles and a rotational isomeric state model. Support is to be found for the crankshaft mechanism in the facts that the activation energy estimated for the model, 13 kcal mole^{-1}, is close to the experimental results, 12–15 kcal mole^{-1}, and that the predicted free volume of activation, about four times the molar volume of a CH_2 unit, is also in good agreement with experimental estimates based on pressure studies.

An outstanding example of the identification of a secondary relaxation with a specific molecular structure lies in the work of Heijboer.[14] Heijboer studied a large number of methacrylate polymers containing the cyclohexyl group in the ester side chain. His results for a series of methyl methacrylate-cyclohexyl methacrylate copolymers are shown in Figure 4-12. It can be seen that the magnitude but not the temperature of the low temperature relaxation is affected by the concentration of cyclohexyl groups. The observed activation energy for this relaxation is 11.5 kcal/mole, identical with that observed for a mechanical relaxation in many low molecular weight cyclohexyl derivatives. This led to the identification of the polymeric relaxation with a flipping of the substituted cyclohexyl ring from one chair conformation to another.

The time-temperature or frequency-temperature superposition scheme discussed in Chapter 3 is applicable to secondary relaxations as well as to the glass transition, assuming that the observed secondary relaxation peaks are well resolved. When the shift factors are

Figure 4-12. Temperature Dependence of tan δ for Copolymers of Methyl Methacrylate and Cyclohexyl Methacrylate. After J. Heijboer, *Kolloid Z.*, **171**, 7 (1960) with the permission of Dr. Dietrich Steinkopff Verlag.

obtained for these secondary relaxations, it is found that they do not obey the WLF equation but an equation of the Arrhenius form, that is

$$\log a_T = -\frac{H_a}{2.303RT} \qquad (4\text{-}26)$$

Thus plots of log a_T versus $1/T$ for a secondary relaxation will yield straight lines, not curves as in the WLF case. This fact has been used to distinguish the main glass transition from other relaxations occurring in semicrystalline polymers.

Problems

1. Assuming that the molecular weight dependence of the fractional free volume is given by

$$f_M = f_\infty + A/\bar{M}_n$$

where f_M is the fractional free volume at molecular weight M, f_∞ is the fractional free volume at infinite molecular weight, and A is a constant, derive equation (4-25).

2.   It has been observed that methacrylate polymers containing linear $(CH_2)_n$ sequences in the ester side chain exhibit a regular decrease in T_g with increasing n. Devise an explanation for this on the basis of the free volume theory.

3.   Qualitatively predict the dependence of T_g on strain on the basis of

  (a) The Free Volume Theory
  (b) The Gibbs-DiMarzio Theory

Might your results suggest an experimental test to distinguish between the two theories? What difficulties might be expected in applying such a test?

4.   (a) If we consider a polymer to be made of only chain ends and chain middles, show that the following equation may be obtained from equation (4-25):

$$T_g = T_g{}^\infty - \frac{cw_e}{M_e}$$

where w_e is the weight fraction of the chain ends and M_e is the weight of chain ends per mole of chains.

  (b) One of the equations used to predict glass transition temperature depression due to the incorporation of solvent is

$$T_g = T_{g_p}w_p + T_{g_s}w_s + Kw_pw_s$$

where subscripts p and s refer to the polymer and the solvent, respectively, and K is a constant. Rearrange this equation into a form similar to the one in part (a) by considering chain ends as solvent.

References

1.   R. S. Porter and J. F. Johnson eds., *Analytical Calorimetry*, Plenum Press, New York, 1969.

2.   P. Ehrenfest, *Leiden Comm. Suppl.*, 756 (1933).

3.   W. Kauzmann, *Chem. Rev.*, **43**, 219 (1948).

4. J. H. Gibbs and E. A. DiMarzio, *J. Chem. Phys.*, **28**, 373 (1958).

5. R. Simha and R. F. Boyer, *J. Chem. Phys.*, **37**, 1003 (1962).

6. L. A. Wood, *J. Polymer Sci.*, **28**, 319 (1958).

7. F. C. Stehling and L. Mandelkern, *Macromol.*, **3**, 242 (1970).

8. M. Gordon and J. S. Taylor, *J. Appl. Chem.*, **2**, 493 (1952).

9. F. E. Karasz and W. J. MacKnight, *Macromol.*, **1**, 537 (1968).

10. R. A. Haldon, W. J. Schell, and R. Simha, *J. Macromol. Sci. Phys.*, **B1**, 759 (1967).

11. W. Lethersich, *Brit. J. Appl. Phys.*, **1**, 294 (1950).

12. N. G. McCrum, B. E. Read, and G. Williams, *Anelastic and Dielectric Effects in Polymeric Solids*, Wiley, New York, 1967.

13. T. F. Schatzki, *J. Polymer Sci.*, **57**, 496 (1962).

14. J. Heijboer, *Kolloid Z.*, **171**, 7 (1960).

5 Statistics of
◆◆◆◆◆◆◆◆◆◆◆◆◆◆◆◆◆◆
a Polymer
◆◆◆◆◆◆◆◆◆◆◆◆◆◆◆◆◆
Chain
◆◆◆◆◆◆◆◆◆◆

The special structure of polymer molecules which distinguishes them from other species is their long, flexible chain nature which has already been discussed in Chapter 1. Such characteristic properties as long-range elasticity, high viscosity, etc., are direct consequences of this special structure. A qualitative overview of the viscoelastic response of polymers and the molecular interpretation of this response has been presented in Chapter 3. In order to provide a quantitative description of the viscoelastic behavior of polymers on the basis of their molecular structure (as opposed to the phenomenological description presented in Chapter 2), we first consider an isolated polymer chain and then extend the results to ensembles of chains, that is, to the bulk polymer itself. An isolated linear polymer chain is capable of assuming many different conformations. Because of thermal agitation, such a chain is also constantly changing from one conformation to another, namely, undergoing Brownian motion. It is, of course, impractical, if not impossible, to analyze all possible conformations in detail. We may only hope to obtain various average properties for the chain by the application of statistical methods.

Due to the presence in real chains of short- and long-range interactions between segments, bond angle restrictions, potential energy barriers, etc., an exact statistical calculation becomes a very complex mathematical problem. However, because of the fact that these chains have such a large number of conformations, idealized models can be used to derive the average properties. In most cases, these quantities provide a good asymptotic approximation to the true values for any sufficiently long linear polymer chain.

We shall first derive the average properties of an ideal polymer chain that is infinitely long, possesses negligibly small volume, and has freely jointed links. Next we shall examine the influence of fixed bond angles between adjacent links. The concept of the statistically equivalent random chain will then be introduced to rationalize the validity of using these model chains to represent the behavior of real polymer chains. Finally, the equation of state for a single polymer chain will be discussed.

A. The Mean Square End-to-End Distance

Suppose our ideal polymer chain has n links, each of length l, then the fully extended length of the chain would be:

$$R = nl \tag{5-1}$$

However, since, as we have just mentioned, the fully extended conformation is only one of a great many, it would be more meaningful to consider an average size of the macromolecule such as the mean square end-to-end distance, $\overline{r^2}$. As the name implies, the end-to-end distance is just the length of the vector connecting the two ends of the ideal chain. The average we seek can be considered to be either the average of the values for a given molecule at a number of times or that of an ensemble of identical molecules at the same time.* Thus, for an ensemble of chains which do not interact with one another

$$\overline{r^2} = (1/p) \sum_{i=1}^{p} \mathbf{r}_i^{\,2} \tag{5-2}$$

---

* For a justification of this assertion, the reader is referred to standard statistics texts such as R. C. Tolman, *Principles of Statistical Mechanics*, Oxford, 1938.

where r_i refers to the end-to-end distance of the ith chain, and p to the number of chains. Now let l_j designate the vectorial length of link j, then

$$\mathbf{r}_i = \sum_{j=1}^{n} \mathbf{l}_j = \mathbf{l}_1 + \mathbf{l}_2 + \mathbf{l}_3 + \cdots + \mathbf{l}_n \tag{5-3}$$

Equation (5-2) now becomes

$$\overline{r^2} = (1/p) \sum_{i=1}^{p} \left(\sum_{j=1}^{n} \mathbf{l}_j \right)_i^2 \tag{5-4}$$

The square of the end-to-end distance of a particular chain is obtained from equation (5-3).

$$r_i^2 = \mathbf{r}_i \cdot \mathbf{r}_i$$
$$= \sum_{j=1}^{n} \mathbf{l}_j \cdot \mathbf{l}_j + 2 \sum_{k<j} \mathbf{l}_k \cdot \mathbf{l}_j \tag{5-5}$$

But from vector algebra we know that the dot product of two vectors is just the product of their absolute values times the cosine of the angle, θ, between one vector and the other vector. Since all the links in our idealized chain are identical in length we have

$$\mathbf{l}_k \cdot \mathbf{l}_j = |\mathbf{l}_k| \, |\mathbf{l}_j| \cos \theta_{kj}$$
$$= l^2 \cos \theta_{kj} \tag{5-6}$$

The angle between the same vectors l_j in the first sum of equation (5-5) is zero, and the cosine of zero is unity. Thus the first sum in equation (5-5) is simply

$$\sum_{j=1}^{n} \mathbf{l}_j \cdot \mathbf{l}_j = nl^2 \tag{5-7}$$

Now in the second sum of equation (5-5) we deal with the product of two different vectors l_k and l_j. We recall that our model chain is freely jointed and volumeless. Thus any two links can assume any orientation whatsoever with respect to each other, including interpenetrating into one another. In other words, any angle θ between two vectors l_k and l_j is just as probable as any other angle. Thus for every $\cos \theta$ in the second term of equation (5-5) there is a $\cos(\pi + \theta) = -\cos \theta$ that exactly cancels it. The result is, of course, that the second term in equation (5-5) is 0. This gives us the mean square end-to-end distance of a freely orienting chain[1-4]:

$$\overline{r^2} = (1/p) \sum_{i=1}^{p} nl^2 \tag{5-8}$$
$$= nl^2$$

In addition, it is possible to calculate the property $\overline{x^2}$, i.e., the average of the square of the projection of $\mathbf{r}_i$ on to any fixed axis, in this case the x axis. Here we may proceed by considering the individual conformations accessible to the chain as before, or we may realize that x^2 is given, in general, as an average over a distribution function

$$\overline{x^2} = \int_0^l x^2 p(x)\, dx \tag{5-9}$$

$p(x)\, dx$ is the probability that x is between x and $x + dx$. To calculate $\overline{x^2}$ we must first determine $p(x)$. The vector $\mathbf{r}_i$ on the average has a length $n^{1/2}l$, and since the entire chain is freely orienting, $\mathbf{r}_i$ itself is evenly distributed through all possible directions. Thus the locus of possible configurations of $\mathbf{r}_i$ is given as the sphere of radius $n^{1/2}l$, one octant of which is shown in Figure 5-1. Now to observe a certain value of x say between x and $x + dx$, it should be clear that $\mathbf{r}_i$ would

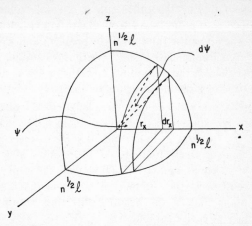

Figure 5-1.   Possible Orientation of **r** in Space.

have to fall in the spherical segment shown. The probability that $\mathbf{r}_i$ will fall in this segment is just equal to the area of this segment, $2\pi(n^{1/2}l)\sin\psi\,d\psi$ divided by the total surface area of the sphere $4\pi(n^{1/2}l)^2$. Thus,

$$p(x)\,dx = 1/2\sin\psi\,d\psi \tag{5-10}$$

and using equation (5-9)

$$\overline{x^2} = \tfrac{1}{2}\int_0^\pi (n^{1/2}l)^2\cos^2\psi\sin\psi\,d\psi$$

$$= \frac{nl^2}{3} = \frac{\overline{r^2}}{3}$$

B. Statistical Distribution of End-to-End Distances

Equation (5-8) tells us that the root mean square length of a freely orienting chain is proportional to the square root of the number of links in the chain. Now we ask: Given a chain of n links, each of which has a length l, what is the probability of achieving a particular end-to-end length? To be more precise, if one end of this chain is at the origin of a Cartesian coordinate system, what are the chances of finding the other end in a small volume element $dx\,dy\,dz$ which is at a distance $\mathbf{r}$ from the origin (Figure 5-2)? This problem is known as the random flight problem and will be solved for the one-dimensional case first.[5-8] We thus constrain $\mathbf{r}$ to lie on the x-axis, then calculate the probability $\omega(x)$ that x has a value between x and $x + dx$. For a chain consisting of a large number of links, we can assume that every link contributes a length $l/\sqrt{3}$ (equation (5-10)) to the component on the x-axis. Some of them will be in the positive x direction (n_+ of them in number) and some in the negative (n_-). Since the chain is freely jointed, the probability of taking either direction will be equal, namely $1/2$. Obviously, the total length of the x-axis component is the number of components in the positive direction minus that in the negative direction multiplied by the average length of each component:

$$x = (n_+ - n_-)l/\sqrt{3} \tag{5-11}$$

Figure 5-2. Conformation of a Polymer Chain with One End Fixed at the Origin of a Cartesian Coordinate System.

Now the probability of a random flight with n_+ positive steps and n_- negative steps is:[14]

$$\omega(n_+, n_-) = (1/2)^n n! / n_+! \, n_-! \tag{5-12}$$

where

$$n = n_+ + n_- \tag{5-13}$$

Let

$$m = n_+ - n_- \tag{5-14}$$

Equation (5-12) becomes

$$\omega(m,n) = (1/2)^n n! \bigg/ \left(\frac{n+m}{2}\right)! \left(\frac{n-m}{2}\right)! \tag{5-15}$$

If $n \gg m$, we can use Stirlings approximation*

$$\ln n! = n \ln n - n \tag{5-16}$$

for $n \to \infty$.

---

* $n! = 1 \times 2 \times 3 \cdots (n-1) \times n$
$\ln n! = \ln 1 + \ln 2 \cdots \ln(n-1) + \ln n = \sum_{x=1}^{n} \ln x$
$\approx \int_1^n \ln x \, dx = [x \ln x - x]_1^n \approx n \ln n - n$

Therefore, after considerable arithmetic rearrangement,

$$\ln[\omega(m,n)] = -\frac{n+m}{2}\ln\left(1+\frac{m}{n}\right) - \frac{n-m}{2}\ln\left(1-\frac{m}{n}\right) \quad (5\text{-}17)$$

However, since $n \gg m$, we can use the following series expansion:

$$\ln\left(1 \pm m/n\right) = \pm m/n - m^2/2n^2 \quad (5\text{-}18)$$

In equation (5-18) third and higher order terms of m/n are neglected. We thus have

$$\ln[\omega(m,n)] = -\frac{m^2}{2n} \quad (5\text{-}19)$$

In using Stirlings approximation some small terms which comprise the normalization of $\omega(m,n)$ have been neglected so that this probability must be renormalized before we can proceed. To do this we may write

$$1 \equiv K \sum_{m=-n}^{m=n} \omega(n,m) \quad (5\text{-}20a)$$

where the probability of m having any value between $-n$ and $+n$ must be unity. We note that the minimum change in Δm is 2, namely, for one bond to change from a positive to a negative orientation or vice versa. Thus in reality equation (5-20a) should be normalized for values of $m = -n, -n+2, -n+4, \ldots, n-2, n$. Mathematically it is simpler to normalize this function over all integer values of m between $-n$ and $+n$ and to realize that this normalization counts twice as many terms and thus should be normalized to 2.0 instead of to 1.0. Thus we rewrite equation (5-20a) as

$$2 \equiv K \sum_{m=-n}^{m=n} \omega(n,m) \quad (5\text{-}20b)$$

where m takes on all integer values. If n is large, equation (5-20b) may be approximated by the integral

$$K \int_{-n}^{n} \omega(n,m)\, dm = 2 \quad (5\text{-}21)$$

Substitution of the nonnormalized equation (5-19) into (5-21) yields upon variable transformation $(y = m/\sqrt{2n})$

$$\sqrt{2n}\,K\int_{-n}^{n} e^{-y^2}\,dy = 2 \qquad (5\text{-}22)$$

If n is sufficiently large

$$\sqrt{2n}\,K\int_{-\infty}^{\infty} e^{-y^2}\,dy = 2 \qquad (5\text{-}23)$$

which is merely

$$\sqrt{2n}\,K\left[2\int_{0}^{\infty} e^{-y^2}\,dy\right] = 2 \qquad (5\text{-}24)$$

The term in brackets in this equation is just the gamma function of argument 1/2 which is equal to $\sqrt{\pi}$. Thus

$$K = \sqrt{\frac{2}{n\pi}} \qquad (5\text{-}25)$$

and the normalized probability $\omega(n,m)$ is given as

$$\omega(n,m) = \sqrt{\frac{2}{n\pi}}\,e^{-m^2/2n} \qquad (5\text{-}26)$$

$\omega(m,n)$ is, however, equal to the probability $\omega(x)\Delta x$ that the length lies between x and $x + \Delta x$

$$\omega(m,n) = \omega(x)\,\Delta x \qquad (5\text{-}27)$$

Once again note that the minimum change in m is 2. Thus Δx corresponds to $\Delta m = 2$. By substituting $\Delta x = 2l/\sqrt{3}$, and $m = \sqrt{3}x/l$ (from equation (5-11)), we get

$$\omega(x)\,dx = (3/2\pi nl^2)^{1/2}\exp(-3x^2/2nl^2)\,dx \qquad (5\text{-}28)$$

Equation (5-28) is known as the Gaussian distribution function. It can be generalized to the three-dimensional case as follows:

$$\omega(x,y,z)\,dx\,dy\,dz = \omega(x)\,dx\,\omega(y)\,dy\,\omega(z)\,dz$$
$$= (b/\pi^{1/2})^3\exp(-b^2r^2)\,dx\,dy\,dz \qquad (5\text{-}29)$$

where

$$b^2 = 3/2nl^2 = 3/2\overline{r^2}$$
$$r^2 = x^2 + y^2 + z^2$$

Equation (5-29) gives the probability that, if one end of a freely orienting chain is fixed at the origin, the other end will be found in the volume element $dx\,dy\,dz$ located at **r** away from the origin.

Rather than the probability of finding the free chain end in a particular volume element $dx\,dy\,dz$, we are often more interested in knowing the probability of finding the free chain end in a spherical shell of radius r centered at the origin. (Here r is not a vector quantity.) In this case, the appropriate function is the so-called radial distribution function $\omega(r)\,dr$ obtained by multiplying $\omega(x,y,z)$ by the volume of a spherical shell of thickness dr located at a distance r from the origin. This volume is $4\pi r^2\,dr$:

$$\omega(r)\,dr = (b/\pi^{1/2})^3 \exp(-b^2 r^2)\,4\pi r^2\,dr \tag{5-30}$$

Equation (5-30) is illustrated in Figure 5-3.

We note that there is a maximum, which corresponds to the most probable end-to-end distance, in the distribution curve. We can find the position of this maximum by differentiating equation (5-30) with respect to r:

$$d\omega(r)/dr = 8\pi r(b/\pi^{1/2})^3\,(1 - b^2 r^2)\exp(-b^2 r^2) \tag{5-31}$$

Setting equation (5-31) equal to zero and solving, the most probable value of r is $1/b$ or $(2nl^2/3)^{1/2}$. The mean square end-to-end distance

Figure 5-3. Radial Distribution Function of a Chain Consisting of 10^4 Freely Orienting Segments each of Length 2.5 Å. After Flory.[12]

is the second moment of the radial distribution function and is given by:

$$\overline{r^2} = \int_0^\infty r^2 \omega(r)\, dr \Big/ \int_0^\infty \omega(r)\, dr \qquad (5\text{-}32)$$

However, the denominator of the right hand side of equation (5-32) is unity since the radial distribution function is normalized. Integration of equation (5-32) yields:

$$\overline{r^2} = \tfrac{3}{2} b^2 = n l^2 \qquad (5\text{-}33)$$

which is just the expression we have derived in the previous section (equation (5-8)).

C.   The Effect of Bond Angle Restrictions

In the preceding two sections, we have treated the conformational statistics of an idealized model chain. In real polymers the situation is, of course, considerably more complicated. However, for the sake of mathematical tractability the quantities derived for the model chain will be used for the development of molecular theories of elasticity and viscoelasticity of polymers to be presented in the later chapters. Nevertheless, it is instructive to consider the effects of incorporating some of the features of a real polymer chain into the model. As an example, we shall treat in this section the effect of bond angle restrictions on the statistics of a polymer chain.

We consider a freely rotating polymer chain whose jth segment orientation is determined by the bond angle it forms with its immediate neighbor, $(j-1)$th segment (Figure 5-4). Combining equations (5-4), (5-5), and (5-7), we have

$$\overline{r^2} = n l^2 + (1/p) \sum_{i=1}^p \left(\sum_{k \neq j}^n \mathbf{l}_k \cdot \mathbf{l}_j \right)_i \qquad (5\text{-}34)$$

However, in this case angles between successive segments cannot assume any value as in the case of the freely orienting chain (no bond angle restrictions) in the preceding section. They must have a fixed

Figure 5-4. Segment of a Freely Rotating Chain with Fixed Bond Angles.

value. The crossterm in equation (5-34) thus will not vanish. To evaluate this term, we consider the dot product $\mathbf{l}_j \cdot \mathbf{l}_{j+1}$ in the sum. The projection of $\mathbf{l}_j$ on its nearest neighbor $\mathbf{l}_{j+1}$ is $l \cos \theta$. Thus $\mathbf{l}_j \cdot \mathbf{l}_{j+1} = l^2 \cos \theta$. The projection of $\mathbf{l}_j$ on its next nearest neighbor $\mathbf{l}_{j+2}$ is then $(l \cos \theta) \cos \theta$ and $\mathbf{l}_j \cdot \mathbf{l}_{j+2} = l^2 \cos^2 \theta$. Therefore equation (5-34) becomes

$$\overline{r^2} = nl^2 + 1/p \sum_{i=1}^{p} [(\cos \theta + \cos^2 \theta + \cos^3 \theta + \cdots + \cos^{n-1} \theta)l^2$$
$$+ (\cos \theta + \cos \theta + \cos^2 \theta + \cdots + \cos^{n-2} \theta)l^2$$
$$+ (\cos^2 \theta + \cos \theta + \cos \theta + \cdots + \cos^{n-3} \theta)l^2$$
$$+ \cdots \cdots \cdots \cdots \cdots \cdots \cdots \cdots \cdots$$
$$+ (\cos^{n-1} \theta + \cos^{n-2} \theta + \cos^{n-3} \theta + \cdots + \cos \theta)l^2]_i \quad (5\text{-}35)$$

Collecting $\cos \theta$ terms, and evaluating the ensemble average, we have

$$\overline{r^2} = nl^2 + [2(n-1) \cos \theta + 2(n-2) \cos^2 \theta$$
$$+ 2(n-3) \cos^3 \theta + \cdots + 2 \cos^{n-1} \theta]l^2$$
$$= nl^2 + 2nl^2 \cos \theta (1 + \cos \theta + \cos^2 \theta + \cdots \quad (5\text{-}36)$$
$$+ \cos^{n-2} \theta) - 2l^2 \cos \theta (1 + 2 \cos \theta + 3 \cos^2 \theta$$
$$+ \cdots + n \cos^{n-3} \theta)$$

The series in equation (5-36) can be summed to yield

$$\overline{r^2} = nl^2 + 2nl^2 \cos\theta(1 - \cos^n\theta)/(1 - \cos\theta)$$
$$- 2l^2 \cos\theta(1 - \cos^n\theta)/(1 - \cos\theta)^2 \qquad (5\text{-}37)$$

The former sum is just a geometric series while the latter series may be generated by the MacLaurin expansion of $(1 - x^n)/(1 - x)^2$. Here we have let $n - 2 \approx n - 1 \approx n$. However since n is so large, the third term on the right hand side of equation (5-37) can be neglected. Cos θ is less than 1, thus $\cos^n\theta$ can also be neglected. Equation (5-37) reduces to

$$\overline{r^2} \approx nl^2[1 + 2\cos\theta/(1 - \cos\theta)] \qquad (5\text{-}38)$$

or

$$\overline{r^2} \approx nl^2(1 + \cos\theta)/(1 - \cos\theta) \qquad (5\text{-}39)$$

For a polymethylene chain, which has tetrahedral bond angles, the angle between a given segment and its projection on the next segment is about 70° (Figure 5-5). Thus, $\cos\theta = 0.34$, and

$$\overline{r^2} = 2nl^2 \qquad (5\text{-}40)$$

which is twice the value of a freely orienting chain.

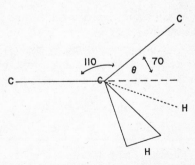

Figure 5-5. Segment of a Polymethylene Chain Showing the Tetrahedral Bond Angles.

D. The Statistically Equivalent Random Chain

We have seen in the preceding sections that drastic approximations have been introduced in our treatment of the statistics of a polymer chain. We must therefore provide ourselves with some justification for representing real chains with these simplified models. A useful concept for this purpose is the statistically equivalent random chain.[9] For a sufficiently long real chain, we can find a random chain with the same mean square end-to-end distance as the real chain.

$$\overline{r^2} = \overline{r_e^2} = n_e l_e^2 \tag{5-41}$$

In equation (5-41) subscripts e refer to the equivalent random chain. However, if both the number and the lengths of segments of the equivalent random chain can be chosen at will, there would be an infinite selection. Thus we must impose another condition; this is chosen so that the fully extended length of the equivalent random chain must be equal to that of the real chain without the distortion of bond angles and lengths:

$$R = R_e = n_e l_e \tag{5-42}$$

Thus there is only one model chain that is statistically equivalent to the real chain with

$$n_e = \overline{R^2}/\overline{r^2}; \qquad l_e = \overline{r^2}/R \tag{5-43}$$

By implication, the real chain then obeys the same Gaussian distribution function as the model chain.

As an example we consider a polymethylene chain (Figure 5-6). Its fully extended length and mean square end-to-end distance are respectively (considering only bond angle restrictions)

$$R = nl \sin(\psi/2) = 0.82nl \tag{5-44}$$

$$\overline{r^2} = 2nl^2 \tag{5-45}$$

Combining with equations (5-43) we get

$$n_e = 0.33n; \qquad l_e = 2.4l \tag{5-46}$$

Figure 5-6. Polymethylene Backbone in the Fully Extended Conformation.

Thus each statistically equivalent random segment is equal to three carbon–carbon bonds in the real polymethylene chain.

The effect of tetrahedral bond angles on the radial distribution function enters through the parameter b, the value of which decreased from $3/2nl^2$ in the freely orienting chain to $3/4nl^2$ in the chain with tetrahedral bond angles.

E. The Equation of State for a Polymer Chain

We have thus far discussed the statistical properties of a single chain which has no external constraints imposed on it. If one end of this polymer chain is fixed at the origin of a coordinate system, then due to Brownian motion the position of the other end of the chain will fluctuate according to the Gaussian distribution function (equation (5-30)). In general there are a large number of conformations which the chain can assume consistent with its ends being separated by a distance r. The number of conformations for each r is proportional to the radial distribution function. If we constrain the chain ends to remain a fixed distance r apart, conformations consistent with all other end-to-end distances become unavailable. As a consequence the "degree of randomness" is now lessened, or in other words, the entropy is decreased. A tension must therefore be set up due to this perturbation. Let us now see how this tension can be related to the chain dimensions and to the magnitude of the Brownian motion which is determined by the thermal energy available to the chain (temperature).

We recall from the first law of thermodynamics that

$$dU = T\,dS - dW \tag{5-47}$$

where U is internal energy, T is temperature, S is entropy, and W is work done by the system. By definition, the Helmholtz free energy is

$$dA = dU - T\,dS \tag{5-48}$$

at constant temperature. Equations (5-47) and (5-48) thus tell us that

$$dA = -dW \tag{5-49}$$

for an isothermal process where the stress-strain work is just

$$dW = -f\,d\mathbf{r} \tag{5-50}$$

Thus the tensile force on a polymer chain at constant temperature and length is just

$$f = -(\partial W/\partial \mathbf{r})_T = +(\partial A/\partial \mathbf{r})_T$$
$$= (\partial U/\partial \mathbf{r})_T - T(\partial S/\partial \mathbf{r})_T \tag{5-51}$$

In the case of our model chain that has no energy barrier hindering the rotation of the segments, the internal energy of the chain is the same for all conformations. Thus the first term on the right hand side of equation (5-51) drops out, and

$$f = -T(\partial S/\partial \mathbf{r})_T \tag{5-52}$$

To calculate the entropy for our Gaussian chain, we use the Boltzmann's relation from statistical mechanics:

$$S = k \ln \Omega \tag{5-53}$$

where k is Boltzmann's constant and Ω is the total number of conformations available to the system. In our case, we want Ω to be a function of $\mathbf{r}$ the vector separation of the chain ends. Let there be $\mathfrak{N}$, some large fixed number of conformations available to a chain. Then the number of conformations consistent with a certain $\mathbf{r}$ is just

$$\Omega(\mathbf{r}) = \mathfrak{N}\omega(\mathbf{r}) \tag{5-54}$$

where $\omega(\mathbf{r})$ is just $\omega(xyz)$ of equation (5-29). Thus,

$$\Omega(\mathbf{r}) = \mathfrak{N}\left(\frac{b}{\pi^{1/2}}\right)^3 e^{-b^2 r^2} \tag{5-55}$$

Inserting equation (5-55) into equation (5-53) and differentiating according to equation (5-52) we obtain the equation of state for a single polymer chain:[10-14]

$$f = 2kTb^2\mathbf{r} \qquad (5\text{-}56)$$

In equation (5-56), f is also a vector. Thus if the ends of a chain are held a fixed distance r apart by a tensile force, then this force is directly proportional to r, and to absolute temperature. It is also inversely proportional to the mean square length of the chain, since $b^2 = 3/2\overline{r^2}$ (equation (5-29)). In fact, equation (5-56) is just Hooke's law for a spring with modulus $2kTb^2$. The elasticity of this spring originates in the decrease in conformational entropy with chain end separation and thus it is often referred to as an "entropy spring." We shall show in Chapter 7 that this entropy spring concept is basic to the formulation of the molecular theory of polymer viscoelasticity. If, instead of a single chain, we have a network of chains, an analogous derivation will give us the equation of state for rubber elasticity, which is the subject of the next chapter.

Problems

1. Write out in detail equation (5-2) with the aid of equations (5-3)–(5-5) if $n = 4$, $p = 1$.

2. The radial distribution function (equation (5-29)) was derived for a chain possessing negligible volume. Suppose now we have a chain whose segments have excluded volume (no two segments can occupy the same space), how would the most probable distance (the maximum in Figure 5-3) be affected?

3. Carry out the integrations indicated in equation (5-32). *Hint:*

$$\int_0^\infty x^{2a}e^{-bx^2}\,dx = \frac{1\cdot 3\cdot 5\cdots(2a-1)}{2^{a+1}b^a}\sqrt{\frac{\pi}{b}}$$

4. Write out in detail equation (5-35) if $n = 4$.

5. It is known that for a poly-*cis*-isoprene chain, each monomer unit is 4.6 Å and $r^2 = 16.2n$. What is the statistically equivalent random chain for this macromolecule?

6. Using the following more exact statement of Stirling's approximation

$$\ln n! = (n + \tfrac{1}{2}) \ln(n - n) + \tfrac{1}{2} \ln 2\pi$$

derive equation (5-26) from equation (5-15) and show that in this case renormalization is not necessary.

7. Calculate $\overline{r^2}$ for a Gaussian chain considering only chain conformations for which $r > \dfrac{1}{2b}$.

8. Calculate $\bar{r}$, the average end-to-end distance for a Gaussian chain.

References

1. H. Eyring, *Phys. Rev.*, **39**, 746 (1932).

2. F. T. Wall, *J. Chem. Phys.*, **11**, 67 (1943).

3. P. Debye, *J. Chem. Phys.*, **14**, 636 (1946).

4. B. H. Zimm and W. H. Stockmayer, *J. Chem. Phys.*, **17**, 1301 (1949).

5. Lord Raleigh, *Phil. Mag.*, **37**, 321 (1919).

6. E. Guth and H. Mark, *Monatsh.*, **65**, 93 (1934).

7. W. Kuhn, *Kolloid-Z.*, **68**, 2 (1934).

8. S. Chandrasekhar, *Rev. Mod. Phys.*, **15**, 3 (1939).

9. W. Kuhn, *Kolloid-Z.*, **76**, 256 (1936); *ibid.*, **87**, 3 (1939).

10. T. M. Birshtein and O. B. Ptitsyn, *Conformations of Macromolecules*, Interscience, New York, 1966.

11. F. Bueche, *Physical Properties of Polymers*, Interscience, New York, 1962, Chap. 1.

12. P. J. Flory, *Principles of Polymer Chemistry*, Cornell University Press, Ithaca, N.Y., 1950, Chap. 10.

13. L. R. G. Treloar, *Physics of Rubber Elasticity*, 2nd ed., Oxford University Press, London, 1958, Chap. 3.

14. M. V. Volkenshtein, *Configurational Statistics of Polymeric Chains*, Interscience, New York, 1963.

6 Rubber
Elasticity

　　　　The high elasticity of rubberlike materials is certainly their most striking characteristic. A crosslinked rubber strip, extended to several times its original length when released will return to that original length exhibiting little or no permanent deformation as a result of the extension. This is, of course, in marked contrast to the behavior of crystalline solids and glasses which cannot normally be extended to more than a very small fraction of their original length without undergoing fracture. It is also in contrast to the behavior of ductile materials such as some metals, which can undergo large deformations without fracture, but which do not return to their original length upon removal of the deforming stress. The elastic response of rubbers has been the subject of a great deal of study by many investigators because of its very great technological importance as well as its intrinsic scientific interest. Starting from one material, namely natural rubber, the development of polymerization techniques has resulted in a host of substances which may properly be called rubbers and a giant synthetic rubber industry has developed to exploit them commercially.

We shall first discuss the thermodynamics of rubber elasticity. The classical thermodynamic approach, as is well known, is only concerned with the macroscopic behavior of the material under investigation and has nothing to do with its molecular structure. The latter belongs to the realm of statistical mechanics, which is the

105

subject of the second section. We then examine the relative significance of energy and entropy in rubber elasticity. A phenomenological treatment then follows, which provides a fuller description of the elasticity at large deformations. Finally, various effects (on rubber elasticity) such as degree of crosslinking, swelling, fillers, and crystallinity are treated.

A. Thermodynamic Treatment

The experimental techniques for investigating the thermodynamic behavior of rubbers can be quite numerous and varied. It is possible to investigate the rubber's response to various kinds of deformations such as shear, tension, compression, etc., and it is also possible to vary temperature, pressure, and volume. Conceptually, one of the easiest experiments to perform is to extend a rubber strip to some fixed length and then to measure the restoring stress exerted by the strip as a function of temperature at constant (atmospheric) pressure. It turns out that such a simple experiment embodies the fundamental thermodynamic principles of rubber elasticity, and thus serves to provide a point of departure for the ensuing development. Since we shall treat rubber elasticity by *equilibrium* thermodynamics, our measurements must be done in accordance with this condition. Experimentally equilibrium is attained by, after stretching, allowing stress-relaxation (see Chapter 3) to proceed for a long time until constant modulus is obtained. Achievement of equilibrium can be evidenced by the reversible changes in elastic force upon changing the temperature.

Figure 6-1 illustrates the equilibrium stress-temperature behavior of natural rubber at various extension ratios.[1] The curves are linear over considerable temperature ranges with negative slopes at low degrees of elongation and positive slopes at higher degrees of elongation. This behavior is not confined to natural rubber but is general for all materials which are called rubbery or rubberlike.

The analysis of the experimental results begins with the combined first and second laws of thermodynamics, applicable to reversible processes. As already shown in the last chapter, it reads

$$dU = T\,dS - dW \qquad (6\text{-}1)$$

Figure 6-1. Stress-Temperature Curves of Natural Rubber. Extension Ratios Are Indicated on the Right of the Figure. After Shen, McQuarrie, and Jackson,[1] with permission of the American Institute of Physics.

In equation (6-1) the increment of work, dW, refers to all of the work performed by the system (i.e., electrical, mechanical, pressure-volume, chemical, etc.). The development of thermodynamics given in most physical chemistry texts is confined to gases where dW becomes simply pressure-volume work, $P\,dV$. In the case of a rubber deformed by an amount dL in tension and exerting a restoring force f, the mechanical work performed on the system to accomplish the deformation, namely $f\,dL$, must also be included in dW. Thus for a rubber strained uniaxially in tension

$$dW = P\,dV - f\,dL \qquad (6\text{-}2)$$

In the measurements discussed above, P is the atmospheric pressure and dV is the volume dilation accompanying the elongation of the rubber. Generally, $P\,dV$ is much smaller than $f\,dL$ and may sometimes be neglected in comparison to $f\,dL$. However, for completeness the treatment following will include pressure-volume work.

Equation (6-1) now becomes

$$dU = T\,dS - P\,dV + f\,dL \qquad (6\text{-}3)$$

Since the experiments discussed were performed under constant pressure conditions, it is natural to use the thermodynamic entity H, the enthalpy. H is defined as

$$H = U + PV \qquad (6\text{-}4)$$

Differentiating equation (6-4) at constant pressure results in

$$dH = dU + P\,dV \qquad (6\text{-}5)$$

Substituting for dU from equation (6-3) we obtain

$$dH = T\,dS + f\,dL \qquad (6\text{-}6)$$

Thus the restoring force exerted by the rubber when it has undergone a deformation by the amount dL at constant temperature and pressure is

$$f = \left(\frac{\partial H}{\partial L}\right)_{T.P} - T\left(\frac{\partial S}{\partial L}\right)_{T.P} \qquad (6\text{-}7)$$

Equation (6-7) shows that the restoring force originates in the enthalpy and entropy changes which occur in the elastomer as a result of the deformation undergone by it.

The change in slope of the stress-temperature curve from negative values at low degrees of elongation to positive values at high degrees of elongation is known as the thermoelastic inversion phenomenon (Figure 6-1). In order to see why this should occur, it is necessary to derive a thermodynamic expression for the slope of this curve, that is an expression for $(\partial f/\partial T)_{P,L}$. This can be done as follows. The Gibbs Free Energy is defined by

$$dF = -S\,dT + V\,dP + f\,dL \qquad (6\text{-}8)$$

One of Maxwell's relations states that

$$\left(\frac{\partial S}{\partial L}\right)_{T,P} = -\left(\frac{\partial f}{\partial T}\right)_{T,L} \tag{6-9}$$

Substitution for $(\partial S/\partial L)_{T,P}$ in equation (6-7) yields

$$f = \left(\frac{\partial H}{\partial L}\right)_{T,P} + T\left(\frac{\partial f}{\partial T}\right)_{P,L} \tag{6-10}$$

Which, upon rearrangement, leads to the desired result:

$$\left(\frac{\partial f}{\partial T}\right)_{P,L} = \frac{f - \left(\dfrac{\partial H}{\partial L}\right)_{T,P}}{T} \tag{6-11}$$

Thus, in order for the slope of the f versus T curves to be negative, it must be that $(\partial H/\partial L)_{T,P} > f$, while in order for the slope to be positive, it must be that $f > (\partial H/\partial L)_{T,P}$. The experimental results show that the former inequality holds at low elongations while the latter inequality holds at high elongations. At sufficiently high elongations it is possible to neglect $(\partial H/\partial L)_{T,P}$ in comparison with f and as a result, under these conditions

$$f = T\left(\frac{\partial f}{\partial T}\right)_{P,L} = -T\left(\frac{\partial S}{\partial L}\right)_{T,P} \tag{6-12}$$

f becomes directly proportional to the absolute temperature, or the elastic response of the rubber is entirely governed by the decrease in entropy which it undergoes upon extension.

Although equation (6-12) adequately describes the behavior of an elastomer at high extensions ($>10\%$ for natural rubber) it is nevertheless true that the coefficient $(\partial H/\partial L)_{T,P}$ has a finite value and cannot be neglected in an exact treatment of rubber elasticity. (It should be stated that equation (6-12) is inadequate to describe the behavior of most elastomers at very high extensions. This is because many elastomers undergo crystallization at sufficiently high extensions and, when this occurs, the term $(\partial H/\partial L)_{T,P}$ once again becomes important and may actually outweigh the term $-T(\partial S/\partial L)_{T,P}$.) In order to explore the origins of the coefficient $(\partial H/\partial L)_{T,P}$, we return to equation (6-5). Differentiation of equation (6-5) at constant temperature yields:

$$\left(\frac{\partial H}{\partial L}\right)_{T,P} = \left(\frac{\partial U}{\partial L}\right)_{T,P} + P\left(\frac{\partial V}{\partial L}\right)_{T,P} \tag{6-13}$$

We wish to separate effects arising from volume changes from effects arising from deformations of the material. To do this, we proceed as follows. Expressing U as a function of length and volume

$$dU = \left(\frac{\partial U}{\partial V}\right)_{L,T} dV + \left(\frac{\partial U}{\partial L}\right)_{V,T} dL \tag{6-14}$$

From equation (6-14)

$$\left(\frac{\partial U}{\partial L}\right)_{T,P} = \left(\frac{\partial U}{\partial V}\right)_{L,T}\left(\frac{\partial V}{\partial L}\right)_{T,P} + \left(\frac{\partial U}{\partial L}\right)_{V,T} \tag{6-15}$$

which, upon substitution into equation (5-13) and rearrangement gives

$$\left(\frac{\partial H}{\partial L}\right)_{T,P} = \left(\frac{\partial U}{\partial L}\right)_{T,V} + \left[\left(\frac{\partial U}{\partial V}\right)_{T,L} + P\right]\left(\frac{\partial V}{\partial L}\right)_{T,P} \tag{6-16}$$

Equation (6-16) shows that the coefficient $(\partial H/\partial L)_{T,P}$ consists of a part arising from internal energy changes occurring during a change of length at constant volume and a volume-dependent part. Although the thermodynamic development is not concerned with the molecular nature of the rubber, the coefficient $(\partial U/\partial L)_{T,V}$ can be related to intramolecular energy effects. This is so because of the constant volume condition. Real rubbers usually have nonzero values for this coefficient.

The coefficient $(\partial U/\partial V)_{T,L}$ may be transformed into experimentally accessible quantities as follows. The Helmholtz Free Energy, dA, is given by

$$dA = -S\,dT - P\,dV + f\,dL \tag{6-17}$$

From this it follows that

$$\left(\frac{\partial S}{\partial V}\right)_{T,L} = \left(\frac{\partial P}{\partial T}\right)_{V,L} \tag{6-18}$$

Returning to equation (6-3) and differentiating with respect to V at constant T and L

$$\left(\frac{\partial U}{\partial V}\right)_{T,L} = T\left(\frac{\partial S}{\partial V}\right)_{T,L} - P \tag{6-19}$$

or, making use of equation (6-18)

$$\left(\frac{\partial U}{\partial V}\right)_{T,L} = T\left(\frac{\partial P}{\partial T}\right)_{V,L} - P \tag{6-20}$$

By a standard technique of partial differentiation

$$\left(\frac{\partial P}{\partial T}\right)_{V,L} = -\left(\frac{\partial P}{\partial V}\right)_{T,L}\left(\frac{\partial V}{\partial T}\right)_{P,L} \tag{6-21}$$

The cubical coefficient of thermal expansion is defined by

$$\alpha = \frac{1}{V}\left(\frac{\partial V}{\partial T}\right)_{P,L} \tag{6-22}$$

and the coefficient of isothermal compressibility by

$$\beta = -\frac{1}{V}\left(\frac{\partial V}{\partial P}\right)_{T,L} \tag{6-23}$$

(Note that, as defined above, both α and β are functions of length inasmuch as the volume of the elastomer undergoes changes on extension.) Using these definitions we have

$$\left(\frac{\partial P}{\partial T}\right)_{V,L} = \frac{\alpha}{\beta} \tag{6-24}$$

and

$$\left(\frac{\partial U}{\partial V}\right)_{T,L} = T\frac{\alpha}{\beta} - P \tag{6-25}$$

Inserting equation (6-25) into equation (6-16)

$$\left(\frac{\partial H}{\partial L}\right)_{T,P} = \left(\frac{\partial U}{\partial L}\right)_{T,V} + T\frac{\alpha}{\beta}\left(\frac{\partial V}{\partial L}\right)_{T,P} \tag{6-26}$$

The quantity $(\partial H/\partial L)_{T,P}$ is directly accessible from experimental f versus T curves as the intercept at $T = 0$ of the tangent to the experimental curve at any desired temperature (see equation (6-10)).

Its measurement thus involves no difficulty for any elastomer provided equilibrium constant pressure stress-temperature data are available for that elastomer. In principle, the second term on the right hand side of equation (6-26) is also directly measurable, but precise values turn out to be very difficult to obtain in practice. Measurements of α and β can be carried out with relative ease, but the coefficient $(\partial V/\partial L)_{T,P}$ is very small for most rubbers and usually depends on extension and thus great experimental skill is required in order to obtain a measurement of any accuracy.[2,3]

By combining equation (6-26) with equation (6-10), the energetic component of the elastic force f_e becomes

$$f_e = \left(\frac{\partial U}{\partial L}\right)_{T,V}$$
$$= f - T\left(\frac{\partial f}{\partial T}\right)_{P,L} - \frac{T\alpha}{\beta}\left(\frac{\partial V}{\partial L}\right)_{T,P} \qquad (6\text{-}27)$$

Thus from equation (6-27) one should be able to compute the relative importance of contributions from energy and entropy to rubber elasticity. Once again, the unavailability of accurate values of the partial $(\partial V/\partial L)_{T,P}$ renders this separation difficult at best.

In general, direct measurements[4] have not been fruitful and various approximate methods have been devised. The latter will be discussed in detail in Section B3 of this chapter.

B. Statistical Treatment

A rubberlike solid is unique in that its physical properties resemble those of solids, liquids, and gases in various respects. It is solidlike in that it maintains dimensional stability, and its elastic response at small strains ($<5\%$) is essentially Hookean. It behaves like a liquid because its coefficient of thermal expansion and isothermal compressibility are of the same order of magnitude as those of liquids. The implication of this is that the intermolecular forces in a rubber are similar to those in liquids. It resembles gases in the sense that the stress in a deformed rubber increases with increasing temperature, much as the pressure in a compressed gas increases with increasing temperature. This gaslike behavior was, in fact, what first provided the hint that rubbery stresses are entropic in origin.

1. DERIVATION

The molecular model for the ideal gas is a collection of point masses in ceaseless, random, thermal motion, the motion of any two of the point masses being completely uncorrelated with one another. The counterpart to this in the case of the ideal rubber is a collection of volumeless, long, flexible chains which are continually undergoing conformational rearrangements due to thermal motion but with all conformations equally accessible and without energy differences among conformations. (That is, all conformations are isoenergetic.) Upon stretching it is assumed that some chain conformations become inaccessible leading to a decrease in entropy and hence to the restoring stress exerted by the rubber.

In order for a rubber to exhibit an equilibrium elastic stress it is necessary for the collection of linear polymer chains to be tied together into an infinite network. Otherwise the Brownian motions of the macromolecules will cause them to move past each other; thus exhibiting flow (Chapter 3). Crosslinking reactions are many and varied. It suffices for our purposes to consider a crosslink to be a permanent tie-point between two chains (Figure 6-2).

To compare the ideal rubber with the ideal gas on a more quantitative basis, we note from equation (6-3) that

$$f = \left(\frac{\partial U}{\partial L}\right)_{T,V} - T\left(\frac{\partial S}{\partial L}\right)_{T,V} \qquad (6\text{-}28)$$

Equation (6-28) resembles the relation

$$-P = \left(\frac{\partial U}{\partial V}\right)_T - T\left(\frac{\partial S}{\partial V}\right)_T \qquad (6\text{-}29)$$

applicable to gases. For ideal gases the internal energy is independent of volume, $(\partial U/\partial V)_T = 0$, and the entropy has two components: one is associated with the heat capacity of the gas, but independent of volume; the other is related to the configurational entropy of the system, and thus a function of the volume. By analogy, the ideal

Figure 6-2. Schematic Diagram of an Ensemble of Linear Polymer Chains Being Crosslinked into an Infinite Network.

rubber may be looked at in the same way. Its internal energy is independent of elongation as required by the model and thus $(\partial U/\partial L)_{T,V} = 0$ and the stress can be attributed to the configurational entropy alone.

The term configurational entropy has been applied historically to the volume-dependent portion of the entropy in an ideal gas and by analogy, the elongation-dependent entropy in the ideal rubber was also referred to as configurational entropy. A difficulty arises with this terminology in rubbers. Rubber molecules are capable of geometrical isomerization, examples of which are cis and trans isomers and stereoisomerism, examples of which are isotactic and syndiotactic forms. These various isomers are known as configurations and cannot be transformed from one into another without breaking chemical bonds. Thus, upon stretching, no configurational changes are possible and there is no configurational contribution to the entropy. What does take place are rotations about single bonds in the chain backbone. It is these conformational changes which give rise to the entropy decrease upon stretching. The conformational statistics of a single chain were the subject of the last chapter. Thus conformational entropy changes upon stretching are the major sources of rubber elasticity, and we shall refer to the entropy involved as conformational entropy in the discussion that follows.

In the formulation of the statistical theory of rubber elasticity, (5-11), the following simplifying assumptions are made:

(1) The internal energy of the system is independent of the conformations of the individual chains.

(2) An individual network chain is freely jointed and volumeless (i.e., it obeys Gaussian statistics (Chapter 5)).

(3) The total number of conformations of an isotropic network of such Gaussian chains is the product of the number of conformations of the individual network chains.

(4) Crosslink junctions in the network are fixed at their mean positions. Upon deformation, these junctions transform affinely, i.e., in the same ratio as the macroscopic deformation ratio of the rubber sample.

Now, from equation (6-17)

$$f = \left(\frac{\partial A}{\partial L}\right)_{T,V} \tag{6-30}$$

But by definition

$$A = U - TS \tag{6-31}$$

Because of assumption (1), however, we need not find an explicit expression for U but can concentrate on the entropy expression. For this we shall again invoke the Boltzmann relation (equation (5-53)), as we did for the isolated chain:

$$S = k \ln \Omega \tag{6-32}$$

where Ω is the total number of conformations available to the rubber network. According to assumption (2), the number of conformations available to the ith individual chain is given by the Gaussian distribution function (equation (5-29)):

$$\omega(x_i,y_i,z_i) = (b/\pi^{1/2})^3 \exp[-b^2(x_i^2 + y_i^2 + z_i^2)] \tag{6-33}$$

Equation (6-33) refers to a chain having an end-to-end vectorial distance $\mathbf{r}_i$ with one end at the coordinates (x_i,y_i,z_i) in the unstrained state, the other end being at the origin of the Cartesian coordinate system (Figure 5-1). Following assumption 3 the total number of conformations available to a network of N such chains is

$$\Omega = \prod_{i=1}^{N} \omega(\mathbf{r}_i) \tag{6-34}$$

and the conformational entropy of the undeformed network is just

$$S_u = 3k \ln (b/\pi^{1/2}) - k \sum_{i=1}^{N} b^2(x_i^2 + y_i^2 + z_i^2) \tag{6-35}$$

or

$$A_u = A_0 + kT \sum_{i=1}^{N} b^2(x_i^2 + y_i^2 + z_i^2) \tag{6-36}$$

where A_0 is that portion of the Helmholtz free energy which is not related to conformational entropy changes.

In the strained state, the chain is deformed to $\mathbf{r}_i'$ with the chain end now at coordinates (x_i',y_i',z_i'). To relate the microscopic strain of the chains to the macroscopic strain of the rubber sample, we assume the deformation to be affine (assumption 4). Consider a unit cube of an isotropic rubber sample (Figure 6-3a). In the general case of a pure homogeneous strain, the cube is transformed into a rectangular parallelepiped (Figure 6-3b). The dimensions of the parallelepiped are λ_1, λ_2, and λ_3 in the three principal axes, where the λ's are called the principal extension ratios. Choosing the coordinate axes for the chain to coincide with the principal axes of strain for the sample, then

$$x_i' = \lambda_1 x_i, \quad y_i' = \lambda_2 y_i, \quad z_i' = \lambda_3 z_i \tag{6-37}$$

The Helmholtz free energy of the deformed network can thus be written as

$$A_d = A_0 + kT \sum_{i=1}^{N} b^2(\lambda_1^2 x_i^2 + \lambda_2^2 y_i^2 + \lambda_3^2 z_i^2) \tag{6-38}$$

The total change in free energy of the rubber network due to the deformation is simply the difference between equations (6-36) and (6-38):

$$\Delta A = kT\left[(\lambda_1^2 - 1) \sum_{i=1}^{N} b^2 x_i^2 + (\lambda_2^2 - 1) \sum_{i=1}^{N} b^2 y_i^2 \right.$$
$$\left. + (\lambda_3^2 - 1) \sum_{i=1}^{N} b^2 z_i^2 \right] \tag{6-39}$$

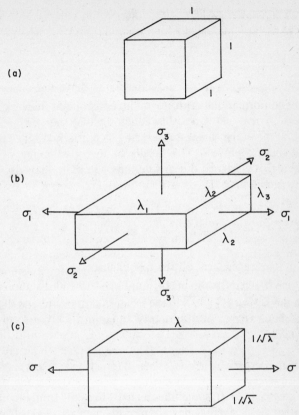

Figure 6-3. A Unit Cube of Rubber: (a) in the Unstrained State; (b) in Homogeneous Strained State; and (c) under Uniaxial Extension.

We recall that by definition:

$$r_i^2 = x_i^2 + y_i^2 + z_i^2 \tag{6-40}$$

For a random, isotropic network all directions are equally probable; then

$$x_i^2 = y_i^2 = z_i^2 = \tfrac{1}{3} r_i^2 \tag{6-41}$$

Equation (6-39) now becomes

$$\Delta A = \tfrac{1}{3} kT \sum_{i=1}^{N} b^2 r_i^2 (\lambda_1^2 + \lambda_2^2 + \lambda_3^2 - 3)$$

$$= \tfrac{1}{3} kNT b^2 \overline{r^2} (\lambda_1^2 + \lambda_2^2 + \lambda_3^2 - 3) \tag{6-42}$$

where we have written $\overline{b^2 r^2} = \sum_{i=1}^{N} b^2 r_i^2 / N$. In Chapter 5, we have defined b^2 for an unstrained freely orienting random chain as

$$b^2 = 3/2\overline{r_f^2} \qquad (6\text{-}43)$$

If the network chains in the unstrained state have the same distribution of conformations as an ensemble of free chains, then $\overline{b^2 r^2} = \frac{3}{2}$. However, in real networks this condition may not be met. For instance, some of the chains may already be partially strained during the crosslinking process. The details of the crosslinking process, e.g., whether it is carried out in solution or in bulk, may also affect the state of the network. It is therefore more general to write the following:

$$\overline{b^2}\,\overline{r^2} = \overline{b^2} \sum_{i=1}^{N} \overline{r_i^2}/N = \frac{\frac{3}{2}(\overline{r_0^2}/\overline{r_f^2})}{N} \qquad (6\text{-}44)$$

where $\bar{b}^2$ is averaged over all the free chains, i.e., $\bar{b}^2 = 3/2\overline{r_f^2}$. Here $\overline{r_0^2} = \sum \overline{r_i^2}/N$ refers to the mean square end-to-end distance of the chain in the network, and $\overline{r_f^2}$ to the mean square end-to-end distance of the isolated chain. Substitution of equation (6-44) into equation (6-42) yields

$$\Delta A = \frac{T}{2} Nk(\overline{r_0^2}/\overline{r_f^2})(\lambda_1^2 + \lambda_2^2 + \lambda_3^2 - 3) \qquad (6\text{-}45)$$

The parameter $(\overline{r_0^2}/\overline{r_f^2})$, sometimes referred to as the front factor, can be regarded as the average deviation of the network chains from the dimensions they would assume if they were isolated and free from all constraints. For an ideal rubber network, the front factor is unity.

Suppose our unit cube has volume V_0 and length L_0. After a uniaxial extension ($\lambda_1 = L/L_0$) is applied, the length in the direction of stretch is L, and the volume dilates to V. We have mentioned previously that the strain-induced volume dilation for unfilled rubbers is very small: of the order of magnitude of 10^{-4}. Nevertheless, the deformation process is not a volume-preserving one, as required by equation (6-30). The device that is generally employed to circumvent this difficulty is to redefine the reference state. A hypothetical hydrostatic pressure is imagined to have been applied to the sample so that its volume in the unstretched state is also V. The initial length is no longer L_0, but is

$$L' = L_0(V/V_0)^{1/3} \tag{6-46}$$

and the extension ratio in the direction of the uniaxial stretch becomes

$$\alpha^* = L/L'$$
$$\alpha_1^* = \lambda_1(V_0/V)^{1/3} \tag{6-47}$$

The average chain dimensions become $\overline{r^2} = \overline{r_0^2}[V/V_0]^{2/3}$. Now the condition of incompressibility requires that

$$\alpha_1^*\alpha_2^*\alpha_3^* = 1 \tag{6-48}$$

while previously

$$\lambda_1\lambda_2\lambda_3 = \frac{V}{V_0} \tag{6-49}$$

Since the network is isotropic the contractions along the two lateral axes are equal and

$$\lambda_1 = \alpha_1^*\left(\frac{V}{V_0}\right)^{1/3}$$
$$\lambda_2 = \lambda_3 = \alpha_1^{*-1/2}\left(\frac{V}{V_0}\right)^{1/3} \tag{6-50}$$

The uniaxial extension of the unit cube is illustrated in Figure 6-3c. Inserting equation (6-50) into equation (6-45) the change in Helmholtz free energy upon deformation becomes

$$\Delta A = \frac{T}{2} Nk(\overline{r^2}/\overline{r_f^2})(\alpha^{*2} + 2/\alpha^* - 3) \tag{6-51}$$

for simple uniaxial extension. Since

$$\left(\frac{\partial A}{\partial L}\right)_{T,V} = \left(\frac{\partial A}{\partial \alpha^*}\right)_{T,V}\left(\frac{\partial \alpha^*}{\partial L}\right)_{T,V} \tag{6-52a}$$

$$f = \frac{T}{L'}\left(\frac{\partial A}{\partial \alpha^*}\right)_{T,V} \tag{6-52b}$$

Performing the indicated differentiation on equation (6-51), we obtain the equation of state for rubber elasticity:

$$f = \frac{NkT}{L'} \left(\frac{\overline{r_0^2}}{\overline{r_f^2}}\right) \left[\alpha^* - \frac{1}{\alpha^{*2}}\right] \left(\frac{V}{V_0}\right)^{2/3} \qquad (6\text{-}53)$$

In equation (6-53) f is the *total* elastic restoring force exerted by the sample. For many purposes it is more convenient to deal with expressions relating the stress to the deformation rather than the total force as in equation (6-53). For this purpose we define a stress: $\sigma = f/A_0$, where A_0 is the cross sectional area of the undeformed sample. We further define $N_0 = N/V_0$ as the number of network chains per unit volume of the undeformed sample, where $V_0 = L_0A_0$. We shall also express N_0 in terms of the number of *moles* of network chains. With the aid of equation (6-47), we can rewrite equation (6-53) as

$$\sigma = N_0RT(\overline{r_0^2}/\overline{r_f^2})(\lambda - V/V_0\lambda^2) \qquad (6\text{-}54)$$

where R is the ideal gas constant ($R = N_{av}k$, N_{av} being the Avogadro's number). The ratio V/V_0 is very nearly unity; thus it is often adequate to write equation (6-54) as

$$\sigma = N_0RT \left(\frac{\overline{r_0^2}}{\overline{r_f^2}}\right)(\lambda - 1/\lambda^2) \qquad (6\text{-}55)$$

The difference between equations (6-54) and (6-55) is numerically trivial, although conceptually important.

The extension ratio may be written

$$\lambda = 1 + \varepsilon \qquad (6\text{-}56)$$

where ε is the tensile strain, $\Delta L/L_0$. By the binomial expansion

$$\lambda^{-2} = (1 + \varepsilon)^{-2} = 1 - 2\varepsilon + \cdots \qquad (6\text{-}57)$$

For very small strains, higher order terms can be neglected and equation (6-55) may be recast as

$$E_0 = \sigma/\varepsilon = 3N_0RT(\overline{r_0^2}/\overline{r_f^2}) \qquad (6\text{-}58)$$

where E_0 is the time-independent tensile modulus. Since, for crosslinked rubbers the tensile modulus is three times the shear modulus G_0 (Chapter 2), then

$$G_0 = N_0 RT(\overline{r_0^2}/\overline{r_f^2}) \tag{6-59}$$

Thus the equation of state can be recast as

$$f = G_0 A_0 \left(\lambda - \frac{1}{\lambda^2}\right) \tag{6-60a}$$

or

$$\sigma = G_0 \left(\lambda - \frac{1}{\lambda^2}\right) \tag{6-60b}$$

where G_0 is determined by the initial slope in the stress-strain curve, and defined by equation (6-59).

The equation of state for rubber elasticity, embodied by any of equations (6-53)–(6-59), is important not only because it is historically the first quantitative treatment of molecular theories for rubbers but also because it laid a conceptual foundation for theories for the physical properties of polymers in general. In later chapters we shall have occasion to discuss some of these in detail. Perhaps the single most significant contribution is its recognition of the role of entropy in polymers, in contradistinction to the predominant role of energy in "ordinary" solids. The equation predicts that the elastic force is directly proportional to temperature and to the total number of chains in the network. Both of these are experimentally observed. The strain dependence of the elastic force is clearly not Hookean, as it is for a single polymer chain (equation (5-56)). Figure 6-4 compares the theory with the experimental stress-strain curve.[12] The agreement is good for strains below 50% or $\lambda < 1.5$, but is poor at high extensions. At these high strains the network chains approach their limiting extensions, and the Gaussian assumption is no longer valid. Another complicating factor is the onset of strain-induced crystallization which we shall discuss further in Section D. Even though the Gaussian theory is only valid at relatively low strains, it is nevertheless extremely valuable in providing a molecular interpretation for rubber elasticity.

2. ENERGY CONTRIBUTION

In our statistical treatment of an ideal rubber, we have assumed that the elastic force is entirely attributable to the conformational

Figure 6-4. Stress-Strain Curve for Natural Rubber. The Theoretical Curve Was Calculated from Equation (6-60) with $G_0 = 4 \times 10^6$ dynes/cm^2. After Treloar[12] with permission of the Faraday Society.

entropy of deformation, energy effects being neglected. That the theory reproduces the essential features of the elasticity of real rubbers attests to the basic soundness of this assumption. On the other hand, we know that in real rubbers such energy effects cannot be entirely absent, and deviations from the ideal rubber model may be expected to occur. Let us now examine in greater detail to what extent the neglect of energy effects is justified.

We can rewrite equation (6-28)

$$f = f_e + f_s \tag{6-61}$$

with

$$f_e = (\partial U / \partial L)_{T,V} \tag{6-62}$$

$$f_s = -T(\partial S / \partial L)_{T,V} \tag{6-63}$$

Since we can experimentally determine f and obtain f_s from the thermodynamic identity (Maxwell relationship resulting from equation (6-17))

$$(\partial S/\partial L)_{T,V} = -(\partial f/\partial T)_{V,L} \qquad (6\text{-}64)$$

We can find f_e by

$$f_e = f - T(\partial f/\partial T)_{V,L} \qquad (6\text{-}65)$$

As discussed in the previous section, the condition of constant volume is difficult to achieve experimentally. Similarly the alternative expression, equation (6-28), also contains terms which are not accurately measurable. We can, however, take advantage of the exact expression of the statistical theory

$$f = G_0 A_0 (\lambda - V/V_0 \lambda^2) \qquad (6\text{-}66)$$

where G_0 is defined by equation (6-59). The shear modulus G_0 is a material constant, which is invariant whether the experimental condition is constant volume or constant pressure. We can therefore differentiate equation (6-66) with respect to T, keeping V and L constant. (Note, however, that V_0 and L_0 are not constant.) The result is

$$\left(\frac{\partial f}{\partial T}\right)_{V,L} = \frac{f \, dG_0}{G_0 \, dT} + \frac{\alpha f}{3} \qquad (6\text{-}67)$$

Combining equations (6-67) and (6-65), we obtain:

$$f_e = f - f \frac{d \ln G_0}{d \ln T} - \alpha T f/3 \qquad (6\text{-}68)$$

The advantage of equation (6-68) is that experimental errors in stress-strain data at various temperatures are averaged out in plotting f against $[\lambda - \lambda^{-2}]$, the slope of which is the shear modulus (Figure 6-5). Knowing shear moduli as a function of temperature, values of f_e can be readily calculated.

For the sake of comparison, it is more meaningful to use the relative energy contribution f_e/f. Thus equation (6-68) can be rewritten as:

$$f_e/f = 1 - \frac{d \ln G_0}{d \ln T} - \frac{\alpha T}{3} \qquad (6\text{-}69)$$

Figure 6-5. Determination of Shear Moduli for Natural Rubber at 10°C and 60°C by Plotting σ Against $\lambda - 1/\lambda^2$. After Shen and Blatz[14] with the permission of the American Institute of Physics.

Values of f_e/f calculated from equation (6-69) are independent of λ, as long as they are obtained within the region of strain for which the Gaussian theory is valid.

Table 6-1 shows the values of f_e/f for several rubbers. We see that in general we cannot expect the contribution of energy to rubber elasticity to be zero. Rather a fraction of the stress is attributable to energy, the rest to entropy. However, since this fraction is a constant as a function of strain, the general *shape* of the stress-strain curve is still unaffected. Thus the neglect of energy effects in the statistical theory essentially causes the predicted curve to differ from the experimental one by a constant fraction. The shear modulus in the statistical theory expression (equation (6-60)) is normally obtained by fitting the experimental stress-strain curve, rather than calculating *a priori* from equation (6-59) (neither N_0 nor $(\overline{r_0^2/r_f^2})$ can at present be determined with certainty by independent measurements). For this reason the energy component of the elastic force is

TABLE 6-1. Thermoelastic Data of Selected Elastomers (Reference temperature: 30°C)

Polymer	$\frac{\alpha}{3} \times 10^4$ (cm/cm/°C)	$d \ln G_0/dT \times 10^3$ (°C^{-1})	f_e/f	Ref.
Poly(ethylene-co-propylene) (EPR)	2.5	2.9	0.04	a
Poly(tetrafluoroethylene-co-perfluoropropylene) (Viton A)	3.0	2.8	0.05	a
Poly(butadiene-co-styrene) (SBR)	2.9	3.4	−0.12	a
Poly(butadiene-co-acrylonitrile) (Hycar)	2.6	2.9	0.03	a
Poly(cis-1,4-butadiene) (Budene)	2.1	2.8	0.10	a
Poly(cis-isoprene) (Natural rubber)	2.2	2.5	0.18	b
Poly(2-hydroxypropyl acrylate)	2.8	3.6	−0.53*	c
Poly(isobutylacrylate)	2.4	3.4	−0.42*	c
Poly(isobutylmethacrylate)	2.2	2.3	0.02*	c

* Reference temperature: 120°C
a E. H. Cirlin, H. M. Gebhard, and M. Shen, *J. Macromol. Sci.*, Part A (1971).
b M. Shen, *Macromol.*, **2**, 358 (1969).
c M. Shen, E. H. Cirlin, and H. N. Gebhard, *Macromol.*, **2**, 682 (1969).

often "absorbed" in the equation of state, rather than separated according to equation (6-61).

We recall that one of the assumptions used in the derivation of the statistical theory is the free energy additivity principle (assumption 3). According to this principle, the number of conformations available to a network of chains is just the product of all the conformations of the individual chains, or the entropy of the network is the sum of the entropies of individual chains in the network. In order for this assumption to be valid, it is required that chains in the network behave as if they were in free space and unaffected by the presence of other chains. This stipulation can only be satisfied if interchain interactions are absent. Thus the energy effects present in rubber elasticity must only come from intrachain interactions, such as the energy barriers hindering rotations along the polymer chain. It is reasonable to expect that real polymer chains are not isoenergetic, i.e., energies of the individual chains are not constant as a function of their conformations. It follows that changes in the supply of thermal energy (changes in temperature) would produce changes in the mean chain dimensions as well. We can derive an expression for the temperature coefficient of the unperturbed dimension of the polymer chain by differentiating equation (6-59) and remembering that $N_0 = N/V_0$ and $\overline{r_0^2}$ is proportional to $V_0^{2/3}$:

$$\frac{d \ln \overline{r_f^2}}{d \ln T} = 1 - \frac{d \ln G_0}{d \ln T} - \frac{\alpha T}{3} \qquad (6\text{-}70)$$

Inserting equation (6-70) into equation (6-69), we see that

$$\frac{d \ln \overline{r_f^2}}{d \ln T} = \frac{f_e}{f} \qquad (6\text{-}71)$$

Thus energy effects in rubber elasticity arise from the intrachain interaction energies of the network chains.

C. Phenomenological Treatment

The statistical theory of rubber elasticity discussed in the preceding section was arrived at through considerations of the underlying molecular dynamics. The equation of state was obtained directly from the Helmholtz free energy of deformation (or simply conformational entropy of deformation, since the energy effects were assumed to be absent), which we can recast with the aid of equations (6-45) and (6-59) as

$$A = -TS = \tfrac{1}{2}G_0(\lambda_1^2 + \lambda_2^2 + \lambda_3^2 - 3) \qquad (6\text{-}72)$$

As we have already seen in Figure 6-4, the theory does not give a complete description of the stress-strain behavior of rubberlike materials. This is perhaps to be expected in view of the many rather drastic simplifying assumptions that have been made in its derivation. That the theory predicts many of the essential features of rubber elasticity is a real tribute to the depth of its insight. However, it is desirable to be able to describe mathematically the stress-strain relations of rubbers over a larger region of strain. For this purpose we must resort to continuum mechanics for a phenomenological treatment of elasticity at large strains.

The phenomenological theory, as its name implies, concerns itself only with the observed behavior of rubbers. It is not based on considerations of the molecular structure of the polymer. The central problem here is to find an expression for the elastic energy stored in the system, analogous to the free energy expression in the statistical theory (equation (6-72)). Consider again the deformation of our unit cube in Figure 6-3. In order to arrive at the state of strain, a certain amount of work must be done which is stored in the body as strain energy:

$$\overline{W}(\lambda_i) = \int_{\lambda_1=1}^{\lambda_1} \sigma_1 \, d\lambda_1 + \int_{\lambda_2=1}^{\lambda_2} \sigma_2 \, d\lambda_2 + \int_{\lambda_3=1}^{\lambda_3} \sigma_3 \, d\lambda_3 \qquad (6\text{-}73)$$

where the λ's are again the principal extension ratios. This energy is a unique function of the state of strain, and if the amount of it is known as a function of strain, the elastic properties of the material can then be completely defined. Although the strain energy function, expressed in terms of the principal extension ratios, is chosen without any regard for the molecular mechanism, it must satisfy certain logical constraints in the case of isotropic solids. These considerations lead to the expression of $\overline{W}$ in terms of the so-called strain invariants:[15,16]

$$\overline{W} = \overline{W}(I_i), \qquad i = 1, 2, 3 \qquad (6\text{-}74)$$

where

$$I_1 = \lambda_1{}^2 + \lambda_2{}^2 + \lambda_3{}^2$$
$$I_2 = \lambda_1{}^2\lambda_2{}^2 + \lambda_2{}^2\lambda_3{}^2 + \lambda_3{}^2\lambda_1{}^2 \qquad (6\text{-}75)$$
$$I_3 = \lambda_1{}^2\lambda_2{}^2\lambda_3{}^2$$

The third strain invariant is obviously

$$I_3 = (V_0/V)^2 \qquad (6\text{-}76)$$

which is equal to unity for an incompressible material.

The most general form of the strain energy function for an

isotropic material is the power series:

$$\overline{W} = \sum_{i,j,k=0}^{\infty} C_{ijk}(I_1 - 3)^i(I_2 - 3)^j(I_3 - 1)^k \qquad (6\text{-}77)$$

Quantities in the parentheses are so chosen that the strain energy vanishes at zero strain. Since we cannot determine *a priori* the set of terms in equation (6-77), let us examine the lowest members of the series. For $i = 1, j = 0, k = 0$:

$$\overline{W} = C_{100}(I_1 - 3) \qquad (6\text{-}78)$$

which is functionally identical to the Gaussian free energy of deformation (equation (6-72)). The stress-strain relation can then be obtained from equation (6-73) by differentiation. Again for the special case of uniaxial extension, using equation (6-50) and setting $C_{100} = C_1$, and observing the incompressibility condition, it follows that

$$\sigma = \partial\overline{W}/\partial\alpha^* = 2C_1(\alpha^* - 1/\alpha^{*2}) \qquad (6\text{-}79)$$

which is known as the neo-Hookean equation. If $C_1 = G_0/2$, equation (6-79) is just the statistical expression (equation (6-60)). Suppose we retain an additional term with $i = 0, j = 1, k = 0$, then

$$\overline{W} = C_{100}(I_1 - 3) + C_{010}(I_2 - 3) \qquad (6\text{-}80)$$

For uniaxial extension, we obtain

$$\sigma = 2C_1(\alpha^* - 1/\alpha^{*2}) + 2C_2(1 - 1/\alpha^{*3}) \qquad (6\text{-}81)$$

where again for simplicity we have set $C_1 = C_{100}$ and $C_2 = C_{010}$. Equation (6-81) is known as the Mooney-Rivlin equation, which can alternatively be expressed as

$$\sigma = 2(C_1 + C_2/\lambda)\,(\lambda - 1/\lambda^2) \qquad (6\text{-}82)$$

since $\lambda = \alpha^*$ in the incompressible case.

The most sensitive way to test the agreement between equation (6-81) and experimental stress-strain data is to plot the quantity $\sigma/(\lambda - 1/\lambda^2)$ against $1/\lambda$. The plot should yield a straight line. Its intercept at $1/\lambda = 0$ is $2C_1$ and its slope is $2C_2$. Figure 6-6 shows a schematic Mooney-Rivlin plot[17] for $C_2/C_1 = 2$. At low strains both $(\lambda - 1/\lambda^2)$ and $1/\lambda$ are extremely sensitive to experimental errors in

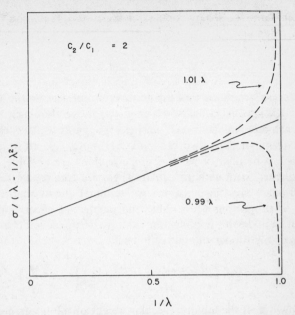

Figure 6-6. Schematic Mooney-Rivlin Plot, Equation (6-82) for $C_2/C_1 =$ 2. Dotted Lines Indicate Curves with 1% Error in λ. Reprinted from *J. Macromol. Sci.—Chem.*, **A1**, 747 (1967). Courtesy of Marcel Dekker, Inc.

extension ratio measurements. Thus if the measured λ is off from the true λ by only 1%, significant deviations from linearity can be expected in this region of strain. The dotted lines in Figure 6-6 illustrate this behavior.

D. Factors Affecting Rubber Elasticity

1. EFFECT OF DEGREE OF CROSSLINKING

According to the statistical theory of rubber elasticity, the elastic stress of a rubber under uniaxial extension is directly proportional to the concentration of network chains N_0. For a rubber sample whose density is d (gm/cc), if the molecular weight of each chain in the network is on the average M_c (gm/mole), then $N_0 = d/M_e$

(mole/cc). Thus the shear modulus of the rubber (equation (6-58)) can be written as

$$G_0 = \frac{dRT}{M_c}\left(\frac{\overline{r_0^2}}{\overline{r_f^2}}\right) \tag{6-83}$$

Equation (6-83) assumes that the network is a perfect one in that all chains in the network are effective in giving rise to the elastic stress. Ideally each crosslink connects four network chains, while each such chain is terminated by two crosslinks. However, as illustrated in Figure 6-2, a number of network imperfections are possible. Each linear polymer chain with molecular weight M, even if all crosslinks are "normal," must give rise to two terminal chains that are incapable of supporting stress. Thus the number of effective chains must not include the imperfections due to chain ends. The shear modulus of the rubber should really be:[18]

$$G_0 = RT\left(\frac{\overline{r_0^2}}{\overline{r_f^2}}\right)\left(\frac{d}{M_c} - \frac{2d}{M}\right) = \frac{dRT}{M_c}\left(\frac{\overline{r_0^2}}{\overline{r_f^2}}\right)\left(1 - \frac{2M_c}{M}\right) \tag{6-84}$$

Note, however, if the initial molecular weight of the linear polymer is infinite, equation (6-84) reduces to equation (6-83). Figure 6-7 shows a plot of G_0 against $1/M$ for a series of natural rubber samples, showing excellent linearity.[19]

Another type of deviation from the "normal" crosslink structure of the rubber is the effect of chain entanglements (Figure 6-2). Such entanglements would impose additional conformational restrictions on the network chains, and thus would have the effect of a "quasi-crosslink" in increasing the elastic stress. Since in actual rubbers the chains are rather closely packed together, one might expect that several such entanglements would occur between the crosslinks. Thus their contribution to stress may be quite significant, especially for chains that are sufficiently long to permit a number of such entanglements. In the absence of an effective way to calculate this quantity, we simply add its contribution to the shear modulus as:

$$G_0 = \left(\frac{dRT}{M_c} + a\right)\left(\frac{\overline{r_0^2}}{\overline{r_f^2}}\right)\left(1 - \frac{2M_c}{M}\right) \tag{6-85}$$

where a represents the entanglement contribution.

Figure 6-7. Variation of Shear Moduli of Natural Rubber with Reciprocal Initial Molecular Weight for Various Vulcanizations. After Mullins[19] with the permission of John Wiley & Sons, Inc.

In addition to terminal chains and entanglements, there are other types of network imperfections. Figure 6-2 shows that if a short chain were crosslinked only once, the crosslink is a wasted one since the chain cannot support elastic stress. Also if a crosslink forms an intrachain loop, it is again an ineffective crosslink. Unfortunately, owing to its very complexity, it is at present impossible to completely characterize the network structure of a rubber.

2. EFFECT OF SWELLING

Linear polymers are capable of dissolving in appropriate solvents to form homogeneous polymer solutions. However, if crosslinks are introduced to tie the chains into an infinite network, the polymer can no longer dissolve. Instead the solvent is absorbed into the polymer network, giving rise to the phenomenon of swelling. A swollen rubber is in fact a solution, except that its mechanical response is now elastic rather than viscous. As solvents fill the network, chains are extended. The resulting retractive force operates in opposition to the swelling force. There is a maximum degree of swelling, at which point these two forces are at equilibrium.

If the rubber is swollen to below the equilibrium swelling so that no de-swelling will occur upon deformation, the statistical expression for the shear modulus (equation (6-59)) can be readily modified. We define V_r as the ratio of the unswollen volume to the swollen one. The number of network chains per unit volume then becomes $N_0 V_r$, and the mean square end-to-end distance of the network chain is now $\overline{r_0^2}/V_r^{2/3}$. Equation (6-59) then reads

$$G_s = N_0 R T V_r^{1/3} (\overline{r_0^2}/\overline{r_f^2})$$
(6-86)

for swollen rubbers. The equation of state is now

$$\sigma_s = N_0 R T V_r^{1/3} (\overline{r_0^2}/\overline{r_f^2})(\lambda_s - 1/\lambda_s^2)$$
(6-87)

where subscripts s refer to the swollen sample. If stress is expressed in terms of per unit cross sectional area of *unswollen* sample, then

$$\sigma_d = N_0 R T V_r^{-1/3} (\overline{r_0^2}/\overline{r_f^2})(\lambda_s - 1/\lambda_s^2)$$
(6-88)

since $A_{0d} = A_{0s} V_r^{2/3}$, where subscripts d refer to the dry (unswollen) sample.

We can also derive an expression on the basis of the Mooney-Rivlin strain energy function for swollen rubbers. A dry rubber sample will undergo two types of deformations: one due to swelling and the other due to extension. The strain energy function per unit volume of swollen rubber is related to that of the dry sample by

$$\overline{W}_s = V_r \overline{W}_d = V_r [C_1(I_1 - 3) + C_2(I_2 - 3)]$$
(6-89)

where subscripts s and d refer to swollen and dry samples respectively. In equation (6-89) the strain invariants I_1 and I_2 are defined by λ_d's, i.e., strains suffered by the dry rubber (both swelling and extension). The deformation due to isotropic swelling is just $V_r^{-1/3}$ for all three principal axes, thus the λ_d's are related to the λ_s's (deformation of swollen rubber by extension) by

$$\lambda_{1d} = \lambda_{1s}V_r^{-\frac{1}{3}}, \quad \lambda_{2d} = \lambda_{2s}V_r^{-\frac{1}{3}}, \quad \lambda_{3d} = \lambda_{3s}V_r^{-\frac{1}{3}} \quad (6\text{-}90)$$

For the case of simple uniaxial extension, assuming incompressibility for the sake of convenience, we obtain with the aid of equations (6-49) and (6-50)

$$\overline{W}_s = C_1 V_r^{\frac{1}{3}}(\lambda_s^2 + 2/\lambda_s - 3) + C_2 V_r^{\frac{5}{3}}(2\lambda_s + 1/\lambda_s^2 - 3) \quad (6\text{-}91)$$

The stress-strain relation can thus be obtained directly by differentiating with respect to λ_s (equation (6-73)):

$$\sigma_s = 2C_1 V_r^{\frac{1}{3}}(\lambda_s - 1/\lambda_s^2) + 2C_2 V_r^{\frac{5}{3}}(1 - 1/\lambda_s^3) \quad (6\text{-}92)$$

or in terms of per unit cross sectional area of unswollen sample:[20]

$$\sigma_d = 2V_r^{-\frac{1}{3}}(C_1 + C_2 V_r^{\frac{1}{3}}/\lambda_s)(\lambda_s - 1/\lambda_s) \quad (6\text{-}93)$$

The C_1 term of the Mooney-Rivlin equation is often identified with the shear modulus of the statistical equation; we see that they both depend on V_r in the same manner (compare equation (6-92) with (6-89) or (6-93) with (6-88)).

For comparison with experimental data, we plot $\sigma_d V_r^{\frac{1}{3}}/2(\lambda_s - 1/\lambda_s^2)$ as a function of $V_r^{\frac{1}{3}}/\lambda_s$. In Figure 6-8 we see that for natural rubbers swollen to various degrees, data all fall on a straight line with the same slope (C_2) and intercept (C_1) in excellent agreement with theory. The upturns in the region of large strains are due to strain-induced crystallization and finite extensibility of the polymer chains, which are not taken into account by either statistical or phenomenological theories.

3. EFFECT OF FILLERS

The use of fillers in rubbers is of paramount technological significance. Such items as automotive tires depend on the addition of fillers to confer on them enhanced wear resistance, strength, and elastic modulus. A wide variety of fillers are commonly employed, such as carbon black, zinc oxide, carbonates and silicates of calcium and magnesium. Generally there are two types of fillers: reinforcing and nonreinforcing. The filler is reinforcing if it is capable of increasing the stiffness of the rubber without impairing its strength and losing its rubbery character.

Figure 6-8. Effect of Swelling on the Mooney-Rivlin Plot of Natural Rubber where V_r is the Volume Fraction of Rubber in the Swollen Sample, (Swelling Liquid, *n*-Decane). After Mullins[21] with the permission of John Wiley & Sons, Inc.

The most commonly used expression for describing the effect of fillers on the elastic modulus of rubbers is the Guth-Smallwood equation:[22,23]

$$E_f/E_0 = 1 + 2.5V_f + 14.1V_f^2 \qquad (6\text{-}94)$$

where V_f is the volume fraction of fillers, and subscripts f and 0 refer to the filled and unfilled rubbers respectively. Equation (6-94) has been found to be valid for a number of filled systems up to a value of V_f of about 0.3. It was actually first derived for the viscosity of liquids with solid suspensions, based on hydrodynamic principles. It was "borrowed" to be used for the elasticity of filled rubbers.

The justification for this is to be found in its good agreement with experimental data,[24] as illustrated in Figure 6-9.

The right hand side of equation (6-94) is sometimes referred to as the strain amplification factor. Suppose we have a sample of rubber in which is dispersed fillers of concentration V_f. The fillers are rigid solid particles with very high modulus. Thus if the filled rubber is deformed, since the fillers are so much more rigid than the rubber, they essentially remain undeformed. Thus all the strain must be suffered by the rubbery phase of the sample. The actual strain sustained by the rubber is greater than the applied strain by the factor given by the right hand side of equation (6-94). Thus we can write the extension ratio appropriate to the rubber matrix as[24]

$$\Lambda = 1 + \varepsilon(1 + 2.5V_f + 14.1V_f^2) \qquad (6\text{-}95)$$

where ε is the applied strain on the filled sample.

Figure 6-10a shows the Mooney-Rivlin plot of a series of natural

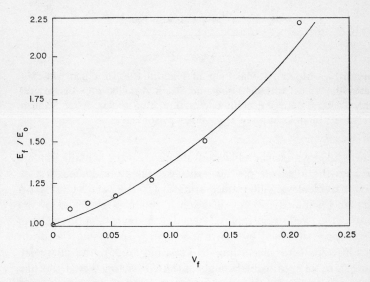

Figure 6-9. Effect of Filler on the Moduli of Natural Rubber Samples (Filler: MT Carbon Black). The Curve Was Calculated from Equation (6-94). After Mullins and Tobin[24] by permission of John Wiley & Sons, Inc.

Figure 6-10. Mooney-Rivlin Plots of Natural Rubber Filled with MT Carbon Black: (a) without Using the Strain Amplification Factor; (b) Using the Strain Amplification Factor, Equation (6-95). After Mullins and Tobin[24] by permission of John Wiley & Sons, Inc.

rubbers filled with carbon black on the basis of the actually applied strain on the filled sample. As expected, the stress increases as a function of increasing filler concentration. The rise at high strains can again be attributed to crystallization or finite extensibility of the chains in the filled sample. Now if we plot the same data on the basis of the strain sustained by the rubber matrix (Λ), using the strain amplification factor according to equation (6-95), the curves in Figure 10b are seen to be brought much closer together. From the slopes and intercepts of these curves, C_1 and C_2 can be readily obtained. The values of these Mooney-Rivlin parameters for the filled samples are quite close to those of the unfilled sample. The agreement is excellent for rubbers filled up to 8%, but poorer for

Figure 6-11. Stress-Softening of Natural Rubber Filled with MPC Carbon Black (Mullins Effect). Numerals Indicate the Stress-Strain Cycles. After F. Bueche, *J. Appl. Polymer Sci.*, **4**, 107 (1960) by permission of John Wiley & Sons.

higher filler contents. These results strongly suggest that the role of fillers is to amplify the applied strain according to equation (6-95). The elastic properties of the rubber itself are unaffected by the presence of these fillers.

Another important effect of fillers is stress-softening, or the Mullins effect. If a filled sample is stretched for the first time to 100%, the stress-strain curve[25] will follow that illustrated in Figure 6-11. Now the strain is removed, and the sample is restretched to 200%. The stress in the second cycle is lower than that in the first up to 100%, after which it continues in a manner following the first cycle. If we repeat the stress-strain in a third cycle, we again see a softening up to 200% due to the previous strain history. This stress-softening effect was first discovered by Mullins, after whom it is named.

A molecular interpretation was first given by F. Bueche[26] for the Mullins effect. Figure 6-12 shows that for a reinforced rubber, the polymer chains are attached to the filler particles. Some of the chains may be relaxed, but some of them are already relatively extended. Upon stretching, the "prestrained" chains will reach maximum extension first and either will become detached from the fillers or

Figure 6-12.   Schematic Diagram of Polymer Chains Attached to Filler Particles. After Bueche.[26]

will be broken. In the second cycle, these broken chains no longer support the stress, thus giving rise to the observed softening in stress. If the stress-strain cycle is repeated for the third time, the same process will be repeated. The softened rubber can be "healed" by annealing at higher temperatures.

4. EFFECT OF STRAIN-INDUCED CRYSTALLIZATION

When a sample of rubber is stretched, it becomes anisotropic in that the network chains tend to orient themselves more in the direction of stretch than in the lateral directions. Thus more chains are ordered to favor the formation of crystallites. These crystallites will tie together a number of neighboring network chains, thereby exerting a crosslinking effect. This increase in the degree of crosslinking will then in turn cause a rise in the elastic stress. The reason that these crystallites in fact act as crosslinks is attributable to their high modulus, which is estimated to be of the order of 10^{12} dynes/cm^2.

This value is some five orders of magnitudes greater than that of most rubbers, which is in the range of 10^7 dyne/cm². If crystallization is allowed to proceed further, more and more of the amorphous material is replaced. At high degrees of crystallinity it is not adequate to regard crystallites just as crosslinks. They will also now act as fillers in further increasing the elastic stress.

That crystallization increases the elastic stress has already been demonstrated in Figure 6-8, in which the Mooney-Rivlin plot shows a rise at high extension ratios. However, it should be remembered that part of this increase is due to finite extensibility of network chains. In Figure 6-13 we show the stress-strain curves of natural rubber at two temperatures.[27] At 0°C there is considerable strain-induced crystallization, and we observe a dramatic rise in the

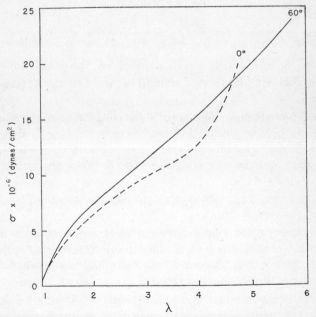

Figure 6-13. Stress-Strain Curves of Natural Rubber at 0°C and 60°C. The Rise in Stress at 0°C at $\lambda > 3$ Is Attributed to Strain-Induced Crystallization. After Smith, Greene, and Ciferri[27] by permission of Dr. Dietrich Steinkopff Verlag.

elastic stress above $\lambda = 3.0$. X-ray measurements show the appearance of crystallinity above this strain. At 60°C there is little or no crystallization, and the stress-strain curve shows a much smaller upturn at high strains. The latter is presumably due only to the finite extensibility of the polymer chains in the network.

Problems

1.   Starting with the equation of state of rubber elasticity (equation (6-60)), show that for an ideal rubber

(a)
$$\left(\frac{\partial H}{\partial L}\right)_{T,P} = \frac{\alpha T}{3}\left[\frac{\lambda^3 + 2}{\lambda^3 - 1}\right]$$

(b)
$$(\partial U/\partial L)_{T,\dot{V}} = 0;$$

(c)
$$(\partial V/\partial L)_{T,P} = \frac{\beta}{3}\left[\frac{\lambda^3 + 2}{\lambda^3 - 1}\right]$$

2.   From the above four equations, find an expression for the stress-temperature coefficient of an ideal rubber at constant pressure and length, i.e., $(\partial f/\partial T)_{P,L}$.

3.   If a rubber sample in the form of a unit cube is deformed by pure shear, then the three principal extension ratios are $\lambda_1 = \lambda$, $\lambda_2 = 1$, $\lambda_3 = 1/\lambda$. (Compare with the case of simple extension in equation (6-50).) Following the arguments of Section B, derive an expression relating σ and λ.

4.   Repeat Problem 3 using the Mooney-Rivlin strain energy function (equation (6-80)).

5.   In analogy to the kinetic theory of ideal gases, the statistical theory of rubber elasticity is often called the kinetic theory of rubber elasticity. Reflect upon the similarities and differences between the basic philosophies of these two theories.

6.   For a piece of ideal rubber, whose density is 0.95 gm/cc, calculate its shear modulus at room temperature if its initial molecular weight is 100,000 gm/mole and the molecular weight of the network chain after crosslinking is 5000 gm/mole (assuming absence of other network defects).

7. What is the maximum degree of swelling for a rubber that would fail at 100% extension in the unswollen state?

8. If the tensile modulus for an unfilled rubber is 5×10^7 dynes/cm^2, what value would it have if it is filled to $V_f = 0.3$? Suppose after several strain cycles, half of its fillers are ineffective due to Mullins effect. What is its tensile modulus now?

9. Starting with equation (6-66), derive equation (6-67).

References

1. M. Shen, D. A. McQuarrie, and J. L. Jackson, *J. Appl. Phys.*, **38**, 791 (1967).

2. G. Gee, J. Stern, and L. R. G. Treloar, *Trans. Faraday Soc.*, **46**, 1101 (1950).

3. F. G. Hewitt and R. L. Anthony, *J. Appl. Phys.*, **29**, 1411 (1958).

4. G. Allen, U. Bianchi, and C. Price, *Trans. Faraday Soc.*, **59**, Part II, 2493 (1963).

5. E. Guth and H. F. Mark, *Monatsh.*, **65**, 93 (1934).

6. W. Kuhn, *Kolloid-Z.*, **68**, 2 (1934); **76**, 258 (1936).

7. F. T. Wall, *J. Chem. Phys.*, **10**, 132, 485 (1942); **11**, 527 (1943).

8. H. M. James and E. Guth, *J. Chem. Phys.*, **11**, 455 (1943); *J. Polymer Sci.*, **4**, 153 (1949).

9. P. J. Flory and J. Rehner, Jr., *J. Chem. Phys.*, **11**, 512, 521 (1943); P. J. Flory, *J. Am. Chem. Soc.*, **18**, 5232 (1956); *Trans. Faraday Soc.*, **56**, 722 (1960).

10. M. S. Green and A. V. Tobolsky, *J. Chem. Phys.*, **14**, 80 (1946); A. V. Tobolsky, D. W. Carlson, and N. Indictor, *J. Polymer Sci.*, **54**, 175 (1961).

11. J. J. Hermans, *Trans. Faraday Soc.*, **43**, 591 (1947); *J. Polymer Sci.*, **59**, 191 (1962).

12. L. R. G. Treloar, *Trans. Faraday Soc.*, **40**, 59 (1944)

13. P. J. Flory, C. A. J. Hoeve, and A. Ciferri, *J. Polymer Sci.*, **34**, 337 (1959).

14. M. Shen and P. J. Blatz, *J. Appl. Phys.*, **39**, 4937 (1968); M. Shen, *Macromol.*, **2**, 358 (1969).

15. M. Mooney, *J. Appl. Phys.*, **11**, 582 (1940).

16. R. S. Rivlin, *Philos. Trans.*, **A241**, 379 (1948).

17. B. M. E. Van der Hoff and E. J. Buckler, *J. Macromol. Sci.*, **A1**, 747 (1967).

18. P. J. Flory, *Chem. Revs.*, **35**, 51 (1944).

19. L. Mullins, *J. Appl. Polymer Sci.*, **2**, 1 (1959).

20. B. M. E. Van der Hoff, *Polymer Sci.*, **6**, 397 (1965).

21. L. Mullins, *J. Appl. Polymer Sci.*, **2**, 257 (1959).

22. E. Guth, R. Simha, and O. Gold, *Kolloid-Z.*, **74**, 266 (1936); E. Guth, *J. Appl. Phys.*, **16**, 20 (1945).

23. H. M. Smallwood, *J. Appl. Phys.*, **15**, 758 (1944).

24. L. Mullins and N. R. Tobin, *J. Appl. Polymer Sci.*, **9**, 2993 (1965).

25. L. Mullins and N. R. Tobin, *Trans. IRI*, **33**, 2 (1956).

26. F. Bueche, *J. Appl. Polymer Sci.*, **4**, 107 (1960); **5**, 27, (1961).

27. K. J. Smith, A. Greene, and A. Ciferri, *Kolloid-Z.*, **194**, 49 (1964).

7 Viscoelastic
◆◆◆◆◆◆◆◆◆◆◆◆◆◆◆◆◆◆◆
Models
◆◆◆◆◆◆◆◆◆◆◆◆

The phenomenological theory of linear viscoelasticity developed in Chapter 2 is completely independent of the existence of models. It is desirable to consider the representation of linear viscoelastic processes by certain model systems in order to gain greater insight into relaxation behavior. In this chapter we consider two broad classes of models. The first consists of the so-called "mechanical analogues." These are combinations of elements, usually springs and dashpots, which more or less faithfully reproduce the viscoelastic response of real systems. The second group is composed of the molecular theories. Here a fairly reasonable representation of a polymer molecule is assumed and the motion of such a molecule in a viscous medium is deduced. In this case the viscoelastic behavior is predicted on the basis of molecular parameters. It will be demonstrated that the two classes of models are equivalent in many respects.

A. Mechanical Elements

We return to the tensile elongation experiment described in the Introduction and Chapter 2. The simplest mechanical model which has some of the gross physical behavior exhibited by bodies subject to uniaxial elongation is a pure Hookean spring (Figure 7-1a). This body is purely elastic and all inertial effects are

143

Figure 7-1. Spring, Dashpot, and Maxwell Model.

neglected. Thus if the Hookean spring is subjected to an instantaneous stress σ_0, it will respond instantaneously with a strain ε_0, σ_0 and ε_0 being related by the equation

$$\sigma_0 = E\varepsilon_0 \qquad (7\text{-}1)$$

The proportionality constant, E, is the Young's modulus with which we are already acquainted. The instantaneous stress application is assumed to produce no oscillations, only a constant strain. It is obvious that no actual substances obey Hooke's law. Some materials, such as steel, obey Hooke's law, as stated in equation (7-1) over a wide range of stress and strain, but none respond without inertial effects.

The dominant characteristic of fluids, on the other hand, is not their elasticity, but rather their viscosity. Newton's law

$$\sigma = \eta \frac{\partial \varepsilon}{\partial t} \qquad (7\text{-}2)$$

is the equation of motion for a model with a simple linear viscous behavior. The mechanical analogue of equation (7-2) is the dashpot element (Figure 7-1b). This is merely a piston in a cylinder filled with a liquid of viscosity η. Integration of equation (7-2) for constant stress σ_0, yields

$$\varepsilon(t) = \frac{\sigma_0}{\eta} t \tag{7-3}$$

Thus subjecting the dashpot to a stress of $10\sigma_0$ for time t produces the same strain as a stress of σ_0 applied for a time $10t$.

1. MAXWELL MODEL

The mechanical response of viscoelastic bodies such as polymers are poorly represented by either the spring or the dashpot. J. C. Maxwell suggested that a better approximation would result from a series combination of the spring and dashpot elements. Such a model is called a Maxwell element and is shown in Figure 7-1c. In the Maxwell element, E, the instantaneous tensile modulus, characterizes the response of the spring while η, the viscosity of the liquid in the dashpot, defines the viscous behavior. One can stipulate a relationship between E and η, without any loss of generality,

$$\eta = \tau E \tag{7-4}$$

τ, the proportionality constant between E and η is known as the relaxation time of the element. The equation of motion of the Maxwell model is

$$\frac{d\varepsilon}{dt} = \frac{1}{E} \frac{d\sigma}{dt} + \frac{\sigma}{\eta} \tag{7-5}$$

Equation (7-5) is merely a linear combination of perfectly elastic behavior as stated in the time derivative of Hooke's law (first term on right hand side) and perfectly viscous behavior as stated in Newton's law (second term on right hand side).

We will now solve this differential equation subject to several sets of experimental boundary conditions.

Creep Experiment. The model is subjected to an instantaneous constant stress σ_0. Thus equation (7-5) becomes

$$\frac{d\varepsilon}{dt} = \frac{\sigma_0}{\eta} \tag{7-6}$$

since $d\sigma/dt$ is zero. Integration of equation (7-6) from time 0 to some time t yields, after division by σ_0

$$\frac{\varepsilon(t)}{\sigma_0} = \frac{\varepsilon_0}{\sigma_0} + \frac{t}{\eta} \qquad (7\text{-}7)$$

One notices that ε_0/σ_0 is just the instantaneous response of the Hookean spring and from equation (7-1) it is seen to be equal to the reciprocal of E. From the results of Chapter 2 we have

$$D = \frac{1}{E} = \frac{\varepsilon_0}{\sigma_0} \qquad (7\text{-}8)$$

where D is the tensile compliance of the spring.

Also from the results of Chapter 2 it is clear that the left hand side of equation (7-7) is simply the tensile creep compliance, $D(t)$. Thus the response of the Maxwell model to a creep experiment is

$$D(t) = D + t/\eta \qquad (7\text{-}9)$$

This is illustrated in Figure 7-2.

Figure 7-2. Maxwell Body Behavior Using Creep Boundary Condition; Log-Log Plot.

Stress Relaxation Experiment. In a stress relaxation experiment, one strains the sample instantaneously to some strain ε_0, and studies the stress $\sigma(t)$ necessary to maintain this constant strain (Chapter 2). The instantaneous strain will be realized only in the spring element. The dashpot will gradually relax so that the stress decreases as a function of time.

After the strain application, $(d\varepsilon/dt)$ is zero so equation (7-5) becomes, in the light of equation (7-4),

$$\frac{d\sigma}{\sigma} = d\ln\sigma = \frac{-dt}{\tau} \tag{7-10}$$

This expression is easily integrated from σ_0 at time 0 to $\sigma(t)$ at time t to give

$$\ln\sigma(t) = \ln\sigma_0 - t/\tau \tag{7-11}$$

Exponentiation and division by ε_0 yields

$$\frac{\sigma(t)}{\varepsilon_0} = \frac{\sigma_0}{\varepsilon_0} e^{-t/\tau} \tag{7-12}$$

Here again, σ_0/ε_0 is the modulus of the spring, E, and the left hand side of equation (7-12) is $E(t)$, the tensile stress relaxation modulus.

$$E(t) = Ee^{-t/\tau} \tag{7-13}$$

Figures 7-3a and 7-3b illustrate the behavior of the Maxwell model in a stress relaxation experiment. It should be noticed that these functions have been plotted on log-log plots as well as the usual rectilinear scales. At time considerably shorter than the relaxation time of the dashpot, the element behaves as if it were a spring alone. At times very long compared to the relaxation time of the dashpot, the model behaves as if it were a dashpot alone, i.e., the stress decays to zero. At times comparable to the relaxation time, the response involves both the spring and the dashpot.

Dynamic Experiments. We will now consider the response of a Maxwell element subjected to a sinusoidal stress. In such a case the strain will also be sinusoidal but out of phase with the stress by the angle δ, as discussed in Chapter 2. Thus

$$\sigma(t) = \sigma_0 e^{i\omega t} \tag{7-14}$$

where σ_0 is the amplitude of the stress and ω is the frequency. Substitution into equation (7-5) yields

$$\frac{d\varepsilon(t)}{dt} = \frac{\sigma_0}{E} i\omega e^{i\omega t} + \frac{\sigma_0}{\eta} e^{i\omega t} \qquad (7\text{-}15)$$

which is easily integrated using the limits of $\varepsilon(t_1)$ at t_1 and $\varepsilon(t_2)$ at t_2. (It should be noted that $\varepsilon(0)$ is not necessarily zero since the stress and strain are not in phase.)

$$\varepsilon(t_2) - \varepsilon(t_1) = \frac{\sigma_0}{E}\left[e^{i\omega t_2} - e^{i\omega t_1}\right] + \frac{\sigma_0}{\eta i\omega}\left[e^{i\omega t_2} - e^{i\omega t_1}\right] \quad (7\text{-}16)$$

Division of the strain increment by the stress increment yields

$$\frac{\varepsilon(t_2) - \varepsilon(t_1)}{\sigma(t_2) - \sigma(t_1)} = D^* = D - i\,\frac{D}{\tau\omega} \qquad (7\text{-}17)$$

Figure 7-3a. Maxwell Body Behavior Using Stress Relaxation Boundary Condition; Linear Plot.

Figure 7-3b. Maxwell Body Behavior Using Stress Relaxation Conditions; Log-Log Plot.

where D^* is the complex tensile creep compliance. In terms of storage and loss compliances (Chapter 2)

$$D^* = D' - iD''$$

$$D' = D \quad \text{and} \quad D'' = \frac{D}{\tau\omega} = \frac{1}{\eta\omega} \qquad (7\text{-}18)$$

From this point, the calculation of the complex tensile modulus, E^*, is straightforward. From Chapter 2, it will be remembered that E^* was shown to be the reciprocal of D^* so we have

$$E^* = \frac{1}{D - \dfrac{iD}{\tau\omega}} = \frac{\tau\omega E}{\tau\omega - i} \qquad (7\text{-}19)$$

Use of the complex conjugate gives

$$E^* = \frac{E\tau^2\omega^2}{1 + \omega^2\tau^2} + \frac{i\tau\omega E}{1 + \omega^2\tau^2} = E' + iE'' \qquad (7\text{-}20)$$

The tangent of the phase angle between the stress and strain was shown to be (Chapter 2)

$$\tan \delta = \frac{E''}{E'} \qquad (2\text{-}21)$$

so

$$\tan \delta = 1/\tau\omega$$
$$E' = E\tau^2\omega^2/(1 + \omega^2\tau^2) \qquad (7\text{-}21)$$
$$E'' = E\tau\omega/(1 + \omega^2\tau^2)$$

for a Maxwell body. Figure 7-4 represents the frequency dependence of E' and E'' for a Maxwell element.

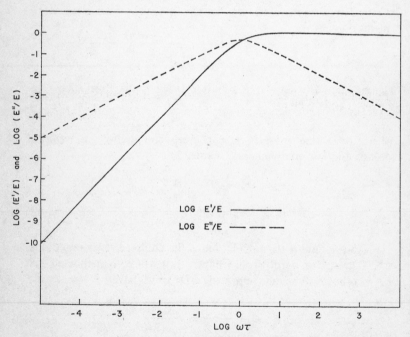

Figure 7-4. Log-Log Plots of $E'(\omega)/E$ and $E''(\omega)/E$ Versus $\omega\tau$ for a Maxwell Model.

2. VOIGT ELEMENT

Another simple element has been used frequently in connection with viscoelastic behavior. The so-called Voigt model consists of the same fundamental elements as the Maxwell model, except here the spring and dashpot are in parallel instead of being in series (Figure 7-5). The constraint on the model is that the strain be the same in both elements. The stress then must be the sum of the stresses in the two individual elements. Thus the fundamental equation of motion for the Voigt element is

$$\sigma(t) = \varepsilon(t)E + \eta \frac{d\varepsilon(t)}{dt} \qquad (7\text{-}22)$$

This model is usually used in considering creep experiments since, as will be seen shortly, it cannot easily be applied to a stress relaxation experiment.

Creep Experiment.   Since the stress is a constant in a creep experiment, one has

$$\frac{d\varepsilon(t)}{dt} + \frac{\varepsilon(t)}{\tau} = \frac{\sigma_0}{\eta} \qquad (7\text{-}23)$$

Equation (7-23) is a linear differential equation which can be made exact and then integrated using the integrating factor $e^{t/\tau}$. Integration

Figure 7-5.   Voigt Model.

between the limits $\varepsilon(0) = 0$ and $\varepsilon(t) = \varepsilon(t)$ yields

$$\int_{\varepsilon(0)}^{\varepsilon(t)} d[\varepsilon(t)e^{t/\tau}] = \frac{\sigma_0}{\eta} \int_0^t e^{t/\tau} \, dt$$

$$\varepsilon(t)e^{t/\tau} = \frac{\sigma_0}{E}(e^{t/\tau} - 1) \tag{7-24}$$

or

$$\frac{\varepsilon(t)}{\sigma_0} = D(t) = D(1 - e^{-t/\tau})$$

Stress Relaxation. With constant strain, the equation of motion of the Voigt element reduces to Hooke's law with

$$E(t) = E \tag{7-25}$$

It is clear, however, that true stress relaxation is an impossible experiment to carry out on a Voigt element since it would take an infinite force to strain the viscous element instantaneously.

The application of sinusoidal stress and strain is similar to that for a Maxwell Body. The results are summarized in Table 7-1 along with the previously derived results for a Maxwell element. Figure 7-6 represents the frequency dependence of D' and D'' for the Voigt element.

The responses of both the Maxwell and Voigt models to several

TABLE 7-1

Experiment	Maxwell Element	Voigt Element
Creep	$D(t) = D + t/\eta$	$D(t) = D(1 - e^{t/\tau})$
Stress relaxation	$E(t) = Ee^{-t/\tau}$	$E(t) = E$
Sinusoidal dynamic experiments	$D' = D$	$D' = \dfrac{D}{1 + \omega^2\tau^2}$
	$D'' = 1/\eta\omega$	$D'' = \dfrac{D\omega\tau}{1 + \omega^2\tau^2}$
	$E' = \dfrac{E\omega^2\tau^2}{1 + \omega^2\tau^2}$	$E' = E$
	$E'' = \dfrac{E\omega\tau}{1 + \omega^2\tau^2}$	$E'' = \omega\eta$

Figure 7-6. Log-Log Plots of $D'(\omega)/D$ and $D''(\omega)/D$ Versus $\omega\tau$ for the Voigt Model.

kinds of deformation experiments are much simpler than those of real polymer systems according to the results presented in Chapter 3. In particular, whereas most linear polymers of sufficiently high molecular weight have at least two major transitions, (glass to rubber and rubber to liquid), these models exhibit only one transition under all conditions. Upon closer examination, it is also apparent that the decay of modulus exhibited by a Maxwell model at times slightly greater than τ is much more rapid than the corresponding modulus decay exhibited by real polymers in either transition region. Thus these simple models do not provide good approximations to the viscoelastic behavior of polymers.

In order to overcome these deficiencies, models have been proposed which consist of combinations of Maxwell and Voigt elements. Although an infinite number of such combinations are possible, we will consider only two here. The treatments of other such models are completely analogous.

3. MAXWELL-WEICHERT MODEL

This generalized model consists of an arbitrary number of Maxwell elements connected in parallel as shown in Figure 7-7.

Consider a Maxwell-Weichert model with z elements subjected to a stress relaxation experiment. The strain in each element is characterized by a spring constant E_i and a viscosity, η_i, thus each τ_i is determined. In all of the individual elements the strain is the same and the total stress σ is the summation of the individual stresses experienced by each element. One can then write

$$\frac{d\varepsilon(t)}{dt} = 0 = \frac{1}{E_1}\frac{d\sigma_1}{dt} + \frac{\sigma_1}{\eta_1}$$

$$= \frac{1}{E_2}\frac{d\sigma_2}{dt} + \frac{\sigma_2}{\eta_2}$$

$$= \frac{1}{E_n}\frac{d\sigma_n}{dt} + \frac{\sigma_n}{\eta_n}$$

$$= \frac{1}{E_z}\frac{d\sigma_z}{dt} + \frac{\sigma_z}{\eta_z} \tag{7-26}$$

$$\sigma = \sigma_1 + \sigma_2 + \sigma_3 + \cdots + \sigma_n + \cdots + \sigma_z \tag{7-27}$$

Figure 7-7.   Maxwell-Weichert Model.

Integration of equation (7-26) gives the partial stresses σ_i which can then be substituted into equation (7-27) to calculate the total stress. When the total stress is divided by the constant strain, ε_0, the stress relaxation modulus results.

$$E(t) = \frac{\sigma(t)}{\varepsilon_0} = \frac{\sigma(0)_1}{\varepsilon_0} e^{-t/\tau_1} + \frac{\sigma(0)_2}{\varepsilon_0} e^{-t/\tau_2}$$

$$+ \cdots + \frac{\sigma(0)_n}{\varepsilon_0} e^{-t/\tau_n} + \cdots + \frac{\sigma(0)_z}{\varepsilon_0} e^{-t/\tau_z}$$

$$= \sum_{i=1}^{z} E_i e^{-t/\tau_i} \qquad (7\text{-}28)$$

where $\sigma(0)_n$ is the stress on the nth element at time equals zero.

Thus we see that the total modulus is the summation of the responses of the individual elements.

The flexibility of this model in reproducing real viscoelastic behavior can be easily demonstrated. Consider a two-element model with parameters

$$E_1 = 3 \times 10^{10}; \ \tau_1 = 1 \quad \text{and} \quad E_2 = 5 \times 10^6; \ \tau_2 = 10^3$$

Its behavior (Figure 7-8) reproduces the two transitions observed in real polymers. It is possible to replace one of the Maxwell elements in the Maxwell-Weichert model with a spring. The stress would decay to a finite value in such a model rather than 0 and would approximate the behavior of crosslinked polymers.

Treatments similar to those used in equations (7-26) and (7-27) can be applied to the Maxwell-Weichert model undergoing sinusoidal stress or strain to calculate the complex modulus. The results are given in Table 7-2. Expressions for the complex compliance are not simple functions.

4. VOIGT-KELVIN MODEL

The Voigt-Kelvin model is a generalization of the Voigt element which results from connecting Voigt elements in series (Figure 7-9). Here the compliance functions are easily calculated, while the modulus functions are rather complicated. The results are summarized in Table 7-2 and a sample calculation is provided below.

Figure 7-8. Behavior of a Two Component Maxwell-Weichert Model in Stress Relaxation.

TABLE 7-2

Experiment	Maxwell-Weichert Model	Voigt-Kelvin Model
Creep		$D(t) = \sum\limits_{i=1}^{z} D_i(1 - e^{-t/\tau_i})$
Stress relaxation	$E(t) = \sum\limits_{i=1}^{z} E_i e^{-t/\tau_i}$	
Sinusoidal dynamic experiments	$E' = \sum\limits_{i=1}^{z} \dfrac{E_i \omega^2 \tau_i^2}{1 + \omega^2 \tau_i^2}$	$D' = \sum\limits_{j=1}^{z} \dfrac{D_i}{1 + \omega^2 \tau_i^2}$
	$E'' = \sum\limits_{i=1}^{z} \dfrac{E_i \omega \tau_i}{1 + \omega^2 \tau_i^2}$	$D'' = \sum\limits_{i=1}^{z} \dfrac{D_i \omega \tau_i}{1 + \omega^2 \tau_i^2}$

Figure 7-9. Voigt-Kelvin Model.

Here we derive expressions for D' and D'' of a Voigt-Kelvin model consisting of z elements assuming a sinusoidal strain application.

Applying equation (7-22) to the Voigt-Kelvin model experiencing a strain in the jth element given by

$$\varepsilon_j(t) = \varepsilon_{0_j} e^{i\omega t}$$

results in the set of equations

$$\sigma(t) = E_j \varepsilon_{0_j} e^{i\omega t} + \eta_j i\omega \varepsilon_{0_j} e^{i\omega t} \quad \text{for} \quad j = 1 \text{ to } z \quad (7\text{-}29)$$

$$\varepsilon(t) = \sum_{j=1}^{z} \varepsilon_j(t) \quad (7\text{-}30)$$

Solving equations (7-29) and (7-30) for the strain in the jth element gives

$$\varepsilon_j(t) = \varepsilon_{0_j}e^{i\omega t} = \frac{\sigma(t)}{E_j + \eta_j i\omega} \qquad (7\text{-}31)$$

which, when substituted into equation (7-30) and simplified, results in

$$\frac{\varepsilon(t)}{\sigma(t)} = D^* = \sum_{j=1}^{z}\frac{1}{E_j + \eta_j i\omega} = \sum_{j=1}^{z}\frac{D_j}{1 + \omega^2\tau_j^2} - i\sum_{j=1}^{z}\frac{D_j\omega\tau_j}{1 + \omega^2\tau_j^2} \quad (7\text{-}32)$$

It is clear that an inversion of equation (7-32) will not yield a simple result.

B. Distributions of Relaxation and Retardation Times

The pertinent parameters used in a Maxwell-Weichert model can be easily presented in graphical form as shown in Figure 7-10. Here three spring constants (10^{10}, 10^8, and 10^6) are associated with relaxation times (10^2, 10^4, and 10^6 respectively). Clearly the

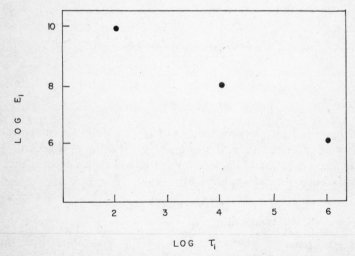

Figure 7-10. Discrete Distribution of Relaxation Times and Associated Partial Modulus Values.

stress relaxation modulus for this particular Maxwell-Weichert model could be easily calculated using equation (7-28).

In considering systems where there are very many Maxwell elements employed in the model, i.e., z in equation (7-28) is large, it is often convenient to replace the summation in the equation by an integration. Thus

$$E(t) = \int_0^\infty E(\tau)e^{-t/\tau}\,d\tau \qquad (7\text{-}33)$$

The various E_i's are replaced by a continuous function, $E(\tau)$, of the relaxation time where this function is called a distribution of relaxation times. In addition to the distribution $E(\tau)$, one often encounters $H(\tau)$ which is simply defined as

$$H(\tau) = \tau E(\tau) \qquad (7\text{-}34)$$

Thus in terms of $H(\tau)$ equation (7-33) becomes

$$E(t) = \int_0^\infty \frac{H(\tau)}{\tau}\, e^{-t/\tau}\,d\tau = \int_{\ln\tau=-\infty}^{\ln\tau=+\infty} H(\tau)e^{-t/\tau}d\ln\tau \qquad (7\text{-}35)$$

Consider calculating the modulus for the distribution of relaxation times given in Figure 7-11. Mathematically $H(\tau)$ can be written as

$$
\begin{aligned}
H(\tau) &= 0 & \log\tau &< 0 \\
H(\tau) &= k\tau^{-1} & 0 &< \log\tau < 1 \\
H(\tau) &= 0 & \log\tau &> 1
\end{aligned}
\qquad (7\text{-}36)
$$

With this $H(\tau)$, equation (7-35) becomes

$$E(t) = \frac{k}{t}\,(e^{-(t/10)} - e^{-t}) \qquad (7\text{-}38)$$

This function is also plotted in Figure (7-11).

In addition dynamic modulus functions can be calculated via this distribution function; $E'(\omega)$ for example is given as

$$E'(\omega) = \int_{\ln\tau=-\infty}^{\ln\tau=\infty} H(\tau)\frac{\omega\tau}{1+\omega^2\tau^2}\,d\ln\tau \qquad (7\text{-}39)$$

A similar method is used to consider compliance functions. Here,

Figure 7-11. Continuous Distribution of Relaxation Times Expressed as $H(\tau)$ and the Corresponding Tensile Stress Relaxation Modulus $E(t)$.

however, the distribution of retardation times $L(\tau)$ defined as

$$J(t) = \int_{\ln \tau = -\infty}^{\ln \tau = \infty} L(\tau)(1 - e^{-t/\tau}) \, d \ln \tau \qquad (7\text{-}40)$$

is used. Although $H(\tau)$ and $L(\tau)$ are related, their exact quantitative relationship is quite complicated and it is suggested that the interested reader consult sources such as Ferry[6] or Gross.[16]

Distributions of relaxation times for real polymer systems are slightly more complex than that shown in Figure 7-11. Tobolsky has suggested, for example, that the stress relaxation modulus of N.B.S. polyisobutylene shown in Figure 7-12 can be thought to arise from a distribution of relaxation times which is composed of a "box and a wedge." This composite is shown in Figure 7-13, and can be expressed as

$$H(\tau) = M/\tau^{\frac{1}{2}} \qquad \tau_1 < \tau < \tau_2$$
$$H(\tau) = E_0 \qquad \tau_3 < \tau < \tau_m$$

In this particular case, Tobolsky gives the pertinent parameters the following values:

$$M = 8.9 \times 10^3 \qquad\qquad E_0 = 7.2 \times 10^5$$
$$\tau_1 = 10^{-12.5} \qquad\qquad \tau_2 = 10^{-5.4}$$
$$\tau_3 = 9.65 \times 10^{-26} \overline{M}_w^{3.30}$$
$$\tau_m = 1.06 \times 10^{-20} \overline{M}_w^{3.30}$$

Figure 7-12. Stress Relaxation Master Curve for N.B.S. Polyisobutylene at 25°C. After A. V. Tobolsky, *Properties and Structure of Polymers*, page 151 with the permission of John Wiley & Sons, Inc.

Figure 7-13. Box and Wedge Distribution of Relaxation Times Equivalent to Master Curve Shown in Figure 7-12. After A. V. Tobolsky, *Properties and Structure of Polymers*, page 128 with the permission of John Wiley & Sons, Inc.

Clearly the wedge is independent of molecular weight and gives rise to the primary transition. The box portion of the spectrum generates the rubbery plateau and rubbery flow regions of the master curve. As discussed in Chapter 3, these regions are strong functions of the molecular weight and the box portion of the spectrum mirrors this fact in the dependence of τ_3 and τ_m on molecular weight.

C. Molecular Theories

Having discussed the viscoelastic responses of simple mechanical models, we may now consider molecular theories. In this treatment it will be shown that the results of molecular theories, can, in fact, be couched in terms of the mechanical models already presented. The molecular theories predict the distribution of relaxation times and partial moduli associated with each relaxation time (τ_i's and E_i's for all i's) which we treated as unknowns or parameters in the previous discussion. Thus, although molecular theories are not based on mechanical models, the results of these treatments may be presented in terms of the parameters of these models. Since, as we have already shown, it is possible to develop expressions giving the

Figure 7-14. Bead Spring Representation of a Real Polymer Molecule in Dilute Solution.

viscoelastic responses of the models to various types of deformations, the predictions of the molecular theories are obtainable through the responses of the models.

Although far from being rigorous, a presentation of some of the salient points of the Rouse theory[1] will be attempted using the method of Peticolas.[2] The conclusions reached in this section are more or less applicable to the theories presented by Rouse,[1] Bueche,[3] and Zimm[4] among others. The aim of this treatment is to present assumptions upon which the theory is based and to outline the results. Mathematical rigor is not attempted in this introductory treatment. For a complete or perhaps accurate understanding of the field, the reader *must* consult the original papers.[1,3,4]

Having considered the treatment of the isolated polymer chain in Chapter 5, one realizes that a long freely orienting molecule behaves like a Hookean entropy spring. (Figure 7-14a.) Equation (5-56) shows that the restoring force on this spring when it is distorted by some amount ΔX is given by

$$f = \frac{3kT}{\overline{r^2}} \Delta X \qquad (5\text{-}56)$$

The molecular theory subdivides the polymer molecule into subunits or submolecules just long enough so that the end-to-end distribution of each of these subunits is Gaussian. This is done so that equation (5-56) is applicable to each submolecule. Our picture of the molecule then becomes much like Figure 7-14b. The mass of the submolecule is concentrated at the beads which are held together by the Hookean springs. Since we will be considering the response of the system to a unidirectional perturbation in the x direction, a spring oriented exactly perpendicular to the x direction will not contribute to the response. Figure 7-15 indicates the deformations allowed for springs in different configurations relative to the perturbing function. It should be clear that an "effective spring constant" in the x direction will now allow us to portray our system as a one-dimensional chain (Figure 7-14c). This linearization of the problem is in no way restrictive. It arises from the facts that the polymer chain is linear and that the deformation considered is a unidirectional one. The only interactions in the model are those of adjacent submolecules with one another.

Figure 7-15. Deformations of Submolecules with Varying Orientation to the Unidirectional Perturbation.

Having proceeded this far, we are now in a position to write the equation of motion for our linear bead and spring model. In order to do this we divide the polymer molecule into z submolecules so that there are z springs and $z + 1$ beads. We now introduce a unidirectional deformation. The restoring force on each of the beads is given by

$$f_{0x} = \frac{-3kT}{a^2}(X_0 - X_1)$$

$$f_{1x} = \frac{-3kT}{a^2}(-X_0 + 2X_1 - X_2)$$

$$f_{ix} = \frac{-3kT}{a^2}(-X_{i-1} + 2X_i - X_{i+1}) \quad 1 \leq i \leq z - 1 \tag{7-42}$$

$$f_{zx} = \frac{-3kT}{a^2}(-X_{z-1} + X_z)$$

In equation (7-42) f_{ix} is the force on the ith bead in the x direction, X_i is the amount by which bead i has been displaced from its equilibrium position and a^2 is the mean square end-to-end distance of the submolecule. The form of equation (7-42) results from the fact that the x directed force on the ith bead results from the difference between the x directed forces on the ith and $i + 1$st segments. An additional force acts on the molecule due to the viscous nature of the medium in which it is immersed. Under the assumption that the beads move in the manner of spheres through a viscous solvent, the drag force on each bead is

$$f_{ix} = \rho \frac{dX_i}{dt} = \rho \dot{X}_i \qquad (7\text{-}43)$$

where ρ is the segmental friction factor. Under the further assumption that the forces arising from the acceleration of the beads are small, the elastic forces given in equation (7-42) and the viscous forces given in equation (7-43) must balance. This allows us to write:

$$\rho \dot{X}_0 = \frac{-3kT}{a^2}(X_0 - X_1)$$

$$\rho \dot{X}_i = \frac{-3kT}{a^2}(-X_{i-1} + 2X_i - X_{i+1}) \qquad 1 \leq i \leq z - 1 \quad (7\text{-}44)$$

$$\rho \dot{X}_z = \frac{-3kT}{a^2}(-X_{z-1} + X_z)$$

This set of linear first order differential equations may be represented according to the matrix notation.

$$[\dot{X}] = -B[A][X]$$

where

$$[\dot{X}] = \begin{bmatrix} \dot{X}_0 \\ \dot{X}_1 \\ \cdot \\ \cdot \\ \cdot \\ \dot{X}_z \end{bmatrix} \qquad [X] = \begin{bmatrix} X_0 \\ X_1 \\ \cdot \\ \cdot \\ \cdot \\ X_z \end{bmatrix}$$

[A] is the square $(z + 1) \times (z + 1)$ matrix

$$[A] = \begin{bmatrix} 1 & -1 & 0 & . & & . & & . & & . \\ -1 & 2 & -1 & 0 & . & & . & & . & . \\ 0 & -1 & 2 & -1 & 0 & & . & & . & . \\ 0 & -1 & . & & & & . & & . & . \\ 0 & -1 & . & & . & . & -1 & & 2 & -1 \\ . & & . & & . & & 0 & & -1 & 1 \end{bmatrix}$$

and B is the composite constant $3kT/a^2\rho$.

Equation (7-44) represents the total physical content of our model. From here on, we will be concerned with mathematical methods which will allow us to solve this apparently complicated set of equations. The main problem arising in any attempted solution of these differential equations results from coupling of the motions of the beads. Thus X_i is not a function of the position of the ith bead itself, but is directly dependent on the position of the adjacent beads. This rather standard problem is effectively treated using the techniques of normal coordinates. We will define a new set of coordinates, q_i, made up of a linear combination of the X_i's. Thus these new coordinates will be defined as

$$q_i = \sum_j Q_{ij} X_j \qquad (7\text{-}45)$$

or in matrix notation

$$[q] = [Q][X]$$

where $[q]$ and $[X]$ are column matrices and $[Q]$ is a $(z + 1)$ by $(z + 1)$ square array. It remains for us to define $[Q]$ in such a way that it will lead to the solution of equation (7-44). To do this we wish to be able to express every equation in the set equation (7-44) in terms of the normal coordinate and its time derivative alone. In matrix notation we formally write:

$$[\dot{q}] = -B[\Lambda][q] \qquad (7\text{-}46)$$

where $[\Lambda]$ unlike $[A]$ is diagonal. Now, for example, the jth equation in the set of equation (7-46) reads

$$\dot{q}_j = -B\lambda_j q_j \qquad (7\text{-}47)$$

Since only the jth normal coordinate and its time derivative appear in equation (7-47), direct integration yields the time dependence of the motion of this coordinate. Clearly, it is in general difficult to have any "feel" for what motion each normal coordinate represents since it has a complicated dependence on *all* of the real coordinates. Nevertheless, the sum of the motion of all of the normal coordinates is identically equal to the sum of the motion of all of the real co-ordinates since one is just a linear transform of the other (equation (7-45)).

Thus to proceed mathematically, we must transform equation (7-44) into equation (7-46) which is done by diagonalizing the matrix $[A]$. There exists another matrix $[Q]$ such that

$$[Q^{-1}][A][Q] = [\Lambda] \tag{7-48}$$

Also

$$[Q^{-1}][Q] = [I] \tag{7-49}$$

In equation (6-37) $[I]$ is a diagonal matrix with all nonzero elements $= 1$. Operation by $[Q^{-1}]$ from the left on equation (7-44) leads to

$$[Q^{-1}][\dot{X}] = -B[Q^{-1}][A][Q][Q^{-1}][X] \tag{7-50}$$

If we now set

$$[Q^{-1}][\dot{X}] = [\dot{q}] \quad \text{and} \quad [Q^{-1}][X] = [q] \tag{7-51}$$

our problem is solved, since equation (7-50) becomes

$$[\dot{q}] = -B[\Lambda][q] \tag{7-52}$$

It must be reemphasized that the exact nature of $[Q^{-1}]$ is not necessary to the physical solution of our problem. Since the normal coordinate approach merely represents a linear transformation of the real coordinates, the motion of the polymer represented by all the q_i's will be identical to the motion of the polymer represented by all the x_i's. Our problem thus becomes the rather simple one of finding a diagonal representation of the $(z + 1) \times (z + 1)$ matrix $[A]$.* This rather well known result (this matrix applies in the treatment of a vibrating string, among others) is derived in Appendix 3, while

---

* It should be noted that $[A]$ was a $(z + 1) \times (z + 1)$ square matrix while $[\Lambda]$ is a $z \times z$. This difference in order represents the fact that while translation of the total molecule is expressed in $[A]$, it is not built into $[\Lambda]$. This causes no complication.

merely being stated here:

$$\lambda_p = 4\sin^2\left(\frac{p\pi}{2(z+1)}\right) \tag{7-53}$$

where p goes from 1 to z. Thus the diagonal matrix $[\Lambda]$ is

$$[\Lambda] = \begin{bmatrix} \lambda_1 & 0 & 0 & . & . & . \\ 0 & \lambda_2 & 0 & . & . & . \\ 0 & 0 & \lambda_3 & . & . & . \\ . & . & . & \lambda_{z+1} & 0 \\ . & . & . & 0 & \lambda_z \end{bmatrix} \tag{7-54}$$

and

$$\dot{q}_p = -B\lambda_p q_p \tag{7-55}$$

which can be directly integrated to give

$$q_p(t) = q_p(0)e^{-Bt\lambda_p} = q_p(0)e^{-t/\tau_p}$$

$$\tau_p = \frac{1}{B\lambda} \qquad p = 1 \text{ to } z \tag{7-56}$$

Here $q_p(0)$ is the value of the normal coordinate at time 0, i.e., at the application of a perturbation and $q(t)$ is the value of the coordinate at time t. We see that the coordinate response is exponential and the system response is just the sum total of all the coordinate responses.

As mentioned above, however, the exact nature of the normal coordinate in terms of the real coordinates is not easily visualizable. Likewise, the exact nature of the real perturbation is not easily visualizable in terms of perturbation to the normal coordinates. Thus to carry out our calculation exactly, one would have to transform the perturbation into the normal coordinate framework. This is exactly the technique used by Bueche in reference 3. The perturbation, i.e., boundary condition, used to solve equation (7-55) was that every normal coordinate was instantaneously displaced to the position of $q_i(0)$ at time zero and then no additional forces were put on the system. This perturbation corresponds neither to creep nor stress relaxation. Although boundary conditions corresponding to these real experiments are more complicated in terms of normal coordinates,[3] it can be shown[2] that the relaxation times which arise in a stress relaxation experiment are just $\frac{1}{2}$ as large as those calculated above. Thus from here on τ_p will be used to denote a relaxation time and is given by

$$\tau_p = \frac{1}{2B\lambda_p} \tag{7-57}$$

Now one may associate those relaxation times with the relaxation times of a Maxwell-Weichert model. Thus the stress relaxation behavior for the bead-spring model is given as

$$E(t) = \sum_{p=1}^{z} E_p e^{-t/\tau_\rho} \tag{7-58}$$

where the individual τ_p's are given in equation (7-57).

It remains to define the individual E_i's and here one normally relies on the kinetic theory of rubber elasticity as applied to the submolecules. Once again

$$f = \frac{3kT}{a^2} \Delta x \tag{7-42}$$

is the elastic force experienced by each spring if the ends are perturbed some amount Δx. Consider however that we have N polymer molecules per cc and the average cross sectional area of each is b^2. Thus the stress experienced by each spring should be given as

$$\sigma = \frac{f}{b^2} = \frac{3kT}{b^2 a^2} \Delta x \tag{7-59}$$

The instantaneous tensile modulus is just the stress divided by the strain so

$$E(0) = \frac{\sigma}{\varepsilon} = \frac{3kT}{b^2 a} \tag{7-60}$$

The denominator of the right hand side of equation (7-60) is merely the volume occupied per submolecule given as

$$ab^2 = \frac{1}{cz} \tag{7-61}$$

Here c is the polymer concentration in molecules per cubic centimeter. Substitution of equation (7-61) into (7-60) yields

$$E(0) = 3ckTz \tag{7-62}$$

Clearly, however, the short time limit of equation (7-58) is just

$$E(0) = \sum_{p=1}^{z} E_p \tag{7-63}$$

Now one usually assumes that all of the individual E_p's are equal and given as

$$E_p = 3ckT \tag{7-64}$$

so that equations (7-62) and (7-63) are consistent. Since we have the τ_p's and E_p's of the Maxwell-Weichert Model we may immediately write, for example,

$$E'(\omega) = 3ckT \sum_{p=1}^{z} \frac{\omega^2 \tau_p^{2}}{1 + \omega^2 \tau_p^{2}} \tag{7-65}$$

D. Applications of Flexible Chain Models to Solutions

Since we are primarily concerned with the viscoelastic nature of bulk materials in this introductory text, we will not dwell on the application of these ideas to solutions but, considering that the model has been derived for the case of a dilute solution, it is of interest to briefly examine its agreement with the observed viscoelastic response of solutions. First it must be recognized that neither a^2, the mean square end-to-end distance of the submolecule nor ρ, the segmental friction factor, may be easily evaluated. Thus a method of eliminating them from our equations will be helpful. The solution viscosity in excess of the solvent viscosity can be thought of as arising from the dissolved polymer and since our polymer behavior has been cast into the Maxwell-Weichert framework, this excess viscosity is just the sum of the viscosity of each of the elements in the model.

$$\eta - \eta_s = \sum_{p=1}^{z} G_p \tau_p \tag{7-66}$$

where η is the viscosity of the solution, η_s is the shear viscosity of the solvent and G_p is one third of E_p. Combining equation (7-66) with (7-64), (7-57), and (7-53) yields,

$$\eta - \eta_s = ckT \sum_{p=1}^{z} \frac{1}{2B\lambda_p} = \frac{ca^2\rho}{24} \sum_{p=1}^{z} \frac{1}{\sin^2 \dfrac{p\pi}{2(z+1)}} \qquad (7\text{-}67)$$

Now for small values of X,

$$\sin X \approx X \qquad (7\text{-}68)$$

and since this expression is in the denominator of each term in the sum in equation (7-67), where the smallest arguments of the sine function will contribute most we may write.

$$\sum_{p=1}^{z} \frac{1}{\sin^2 \dfrac{p\pi}{2(z+1)}} \approx \sum_{p=1}^{z} \frac{4(z+1)^2}{p^2\pi^2} = \frac{4(z+1)^2}{\pi^2} \sum_{p=1}^{z} \frac{1}{p^2} \qquad (7\text{-}69)$$

This summation is well known and for large values of z it is equal to $\pi^2/6$. Furthermore, for large values of z, 1 is small compared to z so that substitution of equation (7-69) into equation (7-67) yields

$$\eta - \eta_s \approx \frac{ca^2\rho z^2}{36} \qquad (7\text{-}70)$$

Applying the same arguments to equation (7-57) yields

$$\tau_p = \frac{a^2\rho z^2}{6\pi^2 kT p^2} \qquad \text{for } p < z \qquad (7\text{-}71)$$

Combination of equation (7-70) and equation (7-71) gives

$$\tau_p = \frac{6(\eta - \eta_s)}{ckT\pi^2 p^2} \qquad (7\text{-}72)$$

where the excess viscosity and concentration of the solution are known or are easily measured.

Rouse and Sittel[5] have investigated the applicability of the theory to real systems, in particular, dilute solutions of polystyrene in the good solvent toluene. Their results are reproduced in Figure 7-16. The agreement between theory and experiment is excellent. However,

Figure 7-16. Experimental Observation and Theoretical Prediction of the Rouse Theory. After J. D. Ferry, *Viscoelastic Properties of Polymers*, Figure 10-4 with the permission of John Wiley & Sons, Inc.

in a sense a certain amount of "curve fitting" is involved since the friction factors and a^2 have been adjusted to fit the data through the method outlined in deriving equation (7-72).

E. The Zimm Modification

Consideration of another major modification which has been applied to the flexible chain model seems pertinent at this point. It has long been appreciated that the velocity gradient of the solvent would be perturbed deep inside a coiled polymer molecule. It is clear that this effect is not considered in the above treatment since the viscous drag is given as $\rho \dot{X}_i$ in equation (7-43) irrespective of whether X_i happens to be inside the coiled molecule or on its surface. Thus one might expect the Rouse formulation to be most applicable to polymer-solvent systems which result in elongated conformations of polymer chains predominating. In such a conformation little shielding from the velocity gradient effective on one part of a molecule by another part of the molecule would result.

This is the case in Figure 7-16 where toluene, a good solvent for polystyrene, would favor extended conformations. Balanced against this is the fact that the model assumes that the Gaussian distribution function holds. This is not true in a solvent which favors extended conformations and renders the agreement shown in Figure 7-16 probably fortuitous.

Zimm[4] has developed a theory which treats such "hydrodynamic shielding" and although we will not go into detail, it is helpful to examine the results of this calculation. The main difference between the Rouse and Zimm treatments occurs in the relaxation times, not in the partial modulus values. The relaxation times according to the Zimm treatment are

$$\tau_p = \frac{1.71(\eta - \eta_s)}{ckTK_p} \tag{7-73}$$

where the K_p are constants; $K_1 = 4.04$; $K_2 = 12.79$; $K_3 = 24.2$ etc. In Figure 7-17 we have compared the prediction of the Rouse and Zimm theories.[6] The limiting values of the slopes of plots of log G' versus log ω at large ω are $\frac{1}{2}$ for the Rouse treatment and $\frac{2}{3}$ for the Zimm treatment. This arises from the different distributions of

Figure 7-17. Predictions of the Rouse and Zimm Theories. After J. D. Ferry et al., *J. Phys. Chem.*, **66**, 536 (1962). Reprinted with the permission of the American Chemical Society.

Figure 7-18. Experimental Results for Polystyrene in Arochlor Compared with Predictions According to the Zimm Theory. After J. D. Ferry et al., *J. Phys. Chem.*, **66**, 536 (1962). Reprinted with the permission of the American Chemical Society.

relaxation times in the Rouse and Zimm theories. In Figure 7-18 the exact results of DeMallie *et al.*,[7] on the system of polystyrene in arochlor are presented. Arochlor, a rather poor solvent for polystyrene, yields a predominance of tightly coiled polymer conformations where hydrodynamic shielding would be expected to be most important. The remarkable agreement between theory and experiment in Figures 7-16 and 7-18 illustrate the validity of the spring and bead model for the viscoelastic behavior of polymer solutions.

F. Extension to Bulk Polymer

Returning to our major concern, bulk polymeric behavior, we may ask if we should expect the bead and spring model to be applicable at all. The equations of motion, equation (7-44) were written on the premise that the beads encountered viscous drag by virtue of their immersion in a solvent with viscosity η. No solvent is present in the bulk case and it would thus appear that equation (7-43) is not applicable to this situation. However, the functional form of the viscous drag term (the left hand side of equation (7-44)) merely states that the solvent medium exerts the same drag on the bead regardless of the beads' position. This functional form can be preserved by substituting an ensemble of bead-spring polymer molecules for the solvent. This procedure has the effect of changing the numerical value of the friction factor, ρ, but retaining the left hand side of equation (7-44) unchanged. The right hand side of equation (7-44) representing the restoring force resulting from the perturbation of a submolecule is an intramolecular property, at least to a first approximation, and thus would not be expected to change drastically with the substitution of additional polymer for solvent. Therefore equation (7-44) is still applicable to the bulk polymer with the appropriate changes in numerical values for such parameters as ρ and a^2. The usual normal coordinate analysis leads directly to equation (7-57) for the distribution of the z relaxation times. The partial modulus now is NkT as opposed to ckT where N is the number of polymer chains per cc of bulk polymeric material. An analysis equivalent to that given in equations (7-66) through (7-72) yields the relationship.

$$\tau_p = \frac{6\eta}{NkT\pi^2 p^2} \tag{7-74}$$

where η now represents the steady flow shear viscosity of a bulk polymer. The time-dependent shear stress relaxation modulus would thus be given as

$$G(t) = NkT \sum_{p=1}^{z} e^{-t/\tau_p} \tag{7-75}$$

A plot of log $G(t)$ versus $\log(t/\tau_1)$ is shown in Figure 7-19. We have assumed that $N = 6 \times 10^{18}$ molecule/cm^3 which corresponds roughly to a molecular weight of 10^5 and a density of 1 gm/cc; NkT, the modulus associated with each relaxation time is then 3.2 × 10^5 dynes/cm^2 at 385°K. The number of terms in the summation

Figure 7-19. Rouse Theory in the Rubbery Plateau and Flow Regions. Varying Numbers of Segments Considered.

of equation (7-75) or equivalently the number of submolecules in the polymer molecule, z, is not determined. We have plotted $E(t)$ for several values of z. It should be clear that $z = 1$ is just the Maxwell body behavior.

Comparison of the curves in this figure with those presented in Chapter 3 giving the observed experimental response of bulk polymers is very disappointing. Experimentally, high molecular weight polymers exhibit two major transitions while the bead-spring model predicts only one.

Ferry, Landel, and Williams[8] have suggested a simple but very appealing explanation for this apparent shortcoming of the theory and, moreover, have put forth a modification of the theory which "predicts" two transitions as observed experimentally. In expounding their treatment, let us revert to some of the considerations regarding the molecular mechanisms giving rise to these transitions as explained in Chapter 3. There we deduced that the behavior in the rubbery region was due mainly to the long range translational motions of an entangled rubbery mass. The molecular motion responsible for the behavior observed in the primary transition region, however, was argued to be of a much shorter range, involving perhaps only a few monomer units. Consider now the segmental friction factor ρ used in equation (7-42). If the motion of submolecule j is part of a coordinated long-range type motion we would expect that the friction factor should reflect the entangled nature of the system since for a long-range motion to take place, entanglements must be affected. Conversely, in short-range motions, entanglements should play only a diminishingly small role; the friction factor would be expected to be indicative of the same chemical species without entanglements (low molecular weight polymer). Thus FLW argued one should expect two independent friction factors to be operative. One of these reflects the viscous drag experienced by a submolecule taking part in short-range uncoordinated motions with short relaxation times and the other indicative of the viscous drag experienced by a submolecule taking part in long-range coordinated motions with long relaxation times.

We now ask which of the two friction factors has been used in the expression for τ_p equation (7-74). ρ has been eliminated from this expression through its dependence on η, the steady state viscosity. Our question becomes then, what molecular motion does η depend upon? Quite possibly it could reflect all motions of the polymer molecules but fortunately for this discussion, it is rather simple to show that only long-range coordinated motions contribute substantially to η. To do this, recall equation (3-7).

$$\eta = \int_0^\infty G(t)\, dt \qquad (3\text{-}7)$$

where we are now calculating a shear viscosity. (See Problem 7-8 for the development of this equation.) A simple transformation yields

$$\eta = \int_{-\infty}^\infty tG(t)\, d\ln t \qquad (7\text{-}76)$$

In Figure 7-20 is plotted the integrand of this integral along with a typically reduced master curve of a high molecular weight polymer. It is evident that the only significant contribution to the area comprising η comes from the long-time portion of the master curve, clearly indicating that the above statement concerning long-range motions as being the sole contribution to η is correct.

Figure 7-20. Calculations of the Viscosity from a Stress Relaxation Master Curve via Equation (3-7).

The friction factor operative for the short-range motions may be thought of as being obtained from measurements on polymers where entanglements are not present and the only contribution to all physical properties, η included, come from short-range motions. Such a polymer is just one of low molecular weight where entanglements are not possible due to the fact that the chains are too short. Polymers with a critical length for the onset of entanglement have been found to have a molecular weight of about 30,000 (polystyrene). Thus FLW postulate one friction factor ρ_0, operative at relaxation times shorter than some critical relaxation time τ_c and another, ρ, operative at longer relaxation times.

$$\tau_p = \frac{\rho_0 a^2 z^2}{6\pi^2 kTp^2} \qquad \tau_p < \tau_c$$

$$\tau_p = \frac{\rho a^2 z^2}{6\pi^2 kTp^2} \qquad \tau_p \geq \tau_c \qquad (7\text{-}77)$$

Moreover, they postulate a relationship ρ/ρ_0 as

$$\log \rho/\rho_0 = 2.4 \log M/M_c \qquad (7\text{-}78)$$

where M_c is the critical molecular weight for the onset of entanglement. We see that they postulate a friction factor ρ which varies as molecular weight to the 2.4 power, unlike ρ_0 which is not a function of molecular weight. The distribution of relaxation times given in equation (7-77) was used to construct the relaxation master curve shown in Figure 7-21 where τ_c was taken to be $\tau_1/400$ and NkT as

Figure 7-21. Two Transitions Predicted Using the Ferry, Landel, and Williams[8] Modification of the Rouse Theory.

1×10^5 dynes/cm^2 with arbitrary ρ/ρ_0. This curve is presented merely to demonstrate that the FLW modification to flexible chain theories does, in fact, straightforwardly predict a two-transition relaxation curve. With a judicious choice of τ_c it is possible to reproduce the gross characteristics of most experimental master curves.

Remembering that the viscosity of a Maxwell-Weichert model is just the sum of the viscosities of the individual Maxwell elements will help shed light on the 2.4 factor in equation (7-78). It has already been noted that all of the viscosity of a high molecular weight polymer derives from long-range translational motion of the polymer, i.e., from motions with $\tau_p > \tau_c$ or where the friction factor ρ is operative. Thus we may write

$$\eta = \sum_{p=1}^{z} G_p \tau_p \qquad \tau_p > \tau_c \qquad (7\text{-}79)$$

and

$$\eta = \frac{NkT\rho a^2 z^2}{6\pi^2 kT} \sum_{p=1}^{p_c} \frac{1}{p^2} \qquad (7\text{-}80)$$

where p_c is defined as

$$p_c = \sqrt{\frac{\tau_1}{\tau_c}} \qquad (7\text{-}81)$$

If p_c is greater than five, for example, the summation is relatively insensitive to changes in p_c and essentially may be considered as a constant. Examination of each term in equation (7-80) with respect to molecular weight dependence reveals

$$\eta \propto M^{3.4} \qquad (7\text{-}82)$$

since N, the number of chains per cubic centimeter is inversely proportional to molecular weight while z, the number of submolecules per polymer molecule, is directly proportional. ρ, it has been pointed out, varies as the 2.4 power of molecular weight while all the other terms are independent of molecular weight. Realizing that equation (7-82) is an experimentally observed fact for high molecular weight materials, makes clear the choice of 2.4 in equation (7-78).

In order to keep this subject in perspective, several shortcomings of the FLW modification of the Rouse theory must be kept in mind. Although these considerations indicate that the modification does

not strictly account for all observations, it must be appreciated that
this was a major step in making a degree of theoretical sense out of a
large amount of sometimes confusing experimental data.

The argument of Williams[9] is perhaps the most serious blow to the
FLW work. He examined the short time limit of equation (7-75) and
observed that

$$G(0) = NkT \sum_{p=1}^{z} (1) = zNkT \tag{7-83}$$

He pointed out that this relationship is true no matter what value of
the friction factor is used since the exponential of $(0/\tau_p)$ is always
1.0. The limiting value of the modulus at short times, which is
experimentally found to be in the neighborhood of 3×10^{10} dynes/
cm^{-2} for most polymers (Chapter 3), is clearly predicted from equation (7-83). Consider a polymer of molecular weight 150,000 and a
density of 1.5 g/cc. The value for NkT at 300°K is 2.5×10^5 dynes/
cm^{-2}. What, then, is the value of z? One should recall that a submolecule was defined as the shortest unit of a chain whose end-to-end
distance is Gaussian. Clearly, this would not be true of a monomeric
or dimeric submolecule due to internal constraints on molecular
geometry. Williams decided that the smallest unit he could accept for
a submolecule was composed of five monomer units. If we let our
polymer be polystyrene whose monomer molecular weight is 100, we
have 1500 monomer units per chain or 300 submolecules per polymer
chain. Using equation (7-83)

$$G(0) = (2.5 \times 10^5 \text{ dynes/cm}^2)(300) = 7.5 \times 10^7 \text{ dynes/cm}^2$$
$$\tag{7-84}$$

which is about two orders of magnitude smaller than the experimentally determined value. The basis for this shortcoming seems to
lie in the fact that one uses NkT as a partial modulus for all chain
motion. This value, while clearly applicable to rubberlike systems,
is not so clearly applicable to deformations of the glassy state.
Tobolsky[10] has pointed out that the equations of motion (equation
(7-43)) can be written and solved for chain units much smaller than
the submolecule while employing a partial modulus more realistic
to the types of molecular motion expected in the glassy state. Thus

a FLW type modification is still feasible in spite of Williams' objection.

Another major discrepancy between theory and experiment is exemplified in Figure 7-22. In this figure are compared the predicted relaxation curves according to the Rouse theory with the experimental results for polystyrene in the primary transition region. It is clear that polystyrene undergoes its glass to rubber transition much more sharply than is predicted by flexible chain theories. Many other polymers, in contrast to this, obey the flexible chain theory result rather closely. Polyisobutylene is one example of this. The discrepancy noted in the case of polystyrene is important, however, as we shall show. The question arises as to what molecular factors contribute to the discrepancy in the case of polystyrene. In order to explore this we use the maximum slope of the stress relaxation master curve in the glass to rubber transition region as a criterion for agreement between theory and experiment. Values of this slope measured experimentally for several polymers are listed in Table 7-3.

The theoretical expression for the modulus is taken from equation (7-75) with the relaxation times given in equation (7-77). Here $\tau_p < \tau_c$ since we are concerned about the transition region where

Figure 7-22. Comparison Between Master Curve of Polystyrene and Predictions of the Modified Rouse Theory in the Primary Transition Region.

TABLE 7-3. Transition Region Slopes for
Several Polymers[12]

Polymer	Slope
Polyvinyl chloride	−0.62
Polyisobutylene	−0.67
Polyethylene terephthalate	−0.87
Polycarbonate	−1.16
Polystyrene	−1.44

short range motions dominate. Let us define a fictitious relaxation time τ_l such that

$$\tau_l = \frac{\rho_0 a^2 z^2}{6\pi^2 kT} \tag{7-85}$$

where τ_l would be the maximum relaxation time τ_1 if one were not forced to change friction factors from ρ_0 to ρ for the long relaxation times. For all $\tau_p < \tau_c$ one may then write

$$\tau_p = \frac{\tau_l}{p^2} \tag{7-86}$$

and equation (7-75) is modified to become

$$G(t) = G \sum_{p=p_c}^{z} e^{-tp^2/\tau_l} \tag{7-87}$$

where G, the partial modulus associated with each relaxation may or may not be NkT but is the same for all short-range molecular motions. Similarly z is the number of molecular units considered per chain and is not necessarily the usual number of submolecules. It is argued now that for relatively large values of p, say $p > 5$, the summation may be replaced by an integration to yield

$$\frac{G(t)}{G} = \int_{p=p_c}^{z} e^{-tp^2/\tau_l} \, dp \tag{7-88}$$

Letting

$$X = \frac{tp^2}{\tau_l}$$

$$\frac{G(t)}{G} = 1/2 \left(\frac{\tau_l}{t}\right)^{\frac{1}{2}} \int_{t/\tau_c}^{t/\tau_{min}} X^{-\frac{1}{2}} e^{-X} \, dX \tag{7-89}$$

where $\tau_{\min} = \tau_l/z^2$. Although this integral cannot be carried out analytically, it is the well-known incomplete gamma function which has been tabulated.[14,15]

$$\Gamma_b(a) = \int_0^b x^{a-1} e^{-x} \, dx \qquad (7\text{-}90)$$

Thus

$$\frac{G(t)}{G} = 1/2 \left(\frac{\tau_l}{t}\right)^{1/2} [\Gamma_{t/\tau_{\min}}(1/2) - \Gamma_{t/\tau_o}(1/2)] \qquad (7\text{-}91)$$

The function $G(t)/G(0)$ has been plotted in Figure 7-23 and, for times between $\tau_{\min}$ and τ_c it is clear that the slope of log $G(t)$ versus log t is $-\frac{1}{2}$; this value arises from the distribution of relaxation times varying as $1/p^2$, the result of the normal mode analysis. Thus we see that this inflexibility of the theory is based in some of the very fundamental aspects of the theory.

Tobolsky[11] has suggested that the response of a highly coupled system would be more like that experimentally observed for polystyrene and has suggested ways to solve the equations generated from these considerations. Instead of considering the response of a linear chain, he treats a network coupled, via springs, in two and

Figure 7-23. Relaxation in the Primary Transition Region Considering Varying Numbers of Relaxing Elements.

three dimensions. This treatment is considered beyond the scope of an introductory book; interested individuals are referred to the original papers.[10–13]

Appendix

Equation (7-53), which will be derived in this appendix, represents the eigenvalues used in diagonalizing the matrix $[A]$ of equation (7-44)

$$[\dot{X}] = -B[A][X] \qquad (7\text{-}44)$$

where $[A]$ is a square matrix of order $(z + 1)$.

$$[A] = \begin{bmatrix} 1 & -1 & 0 & 0 & \cdots\cdots\cdots \\ -1 & 2 & -1 & 0 & \cdots\cdots\cdots \\ 0 & -1 & 2 & -1 & \cdots\cdots\cdots \\ \cdots\cdots\cdots\cdots\cdots\cdots\cdots\cdots \\ \cdots\cdots\cdots\cdots & -1 & 2 & -1 \\ \cdots\cdots\cdots\cdots & 0 & -1 & 1 \end{bmatrix}$$

First transform $[A]$ into a form which will be easier to treat. To do this, note that

$$[A] = [C^T][C]$$

where $[C]$ is also a matrix of order $z \times (z + 1)$ and has the form

$$[C] = \begin{bmatrix} 1 & -1 & 0 & 0 & \cdots\cdots\cdots \\ 0 & 1 & -1 & 0 & \cdots\cdots\cdots \\ 0 & 0 & 1 & -1 & \cdots\cdots\cdots \\ \cdots\cdots\cdots\cdots\cdots\cdots\cdots\cdots \\ \cdots\cdots\cdots\cdots & 1 & -1 & 0 \\ \cdots\cdots\cdots\cdots & 0 & 1 & -1 \end{bmatrix}$$

and

$$[C^T] = \begin{bmatrix} 1 & 0 & 0 & 0 & \cdots\cdots\cdots \\ -1 & 1 & 0 & 0 & \cdots\cdots\cdots \\ 0 & -1 & 1 & 0 & \cdots\cdots\cdots \\ \cdots\cdots\cdots\cdots\cdots\cdots\cdots\cdots \\ \cdots\cdots\cdots\cdots\cdots & -1 & 1 \\ \cdots\cdots\cdots\cdots\cdots & 0 & -1 \end{bmatrix}$$

Thus, equation (7-44) can be rewritten as

$$[\dot{X}] = -B[C^T][C][X] \tag{a}$$

Now premultiply both sides of equation (a) by $[C]$:

$$[C][\dot{X}] = -B[C][C^T][C][X]$$
$$= -B[R][C][X] \tag{b}$$

Where $[R]$ is called the Rouse matrix,

$$[R] = \begin{bmatrix} 2 & -1 & 0 & 0 & \cdots\cdots\cdots\cdots\cdots \\ -1 & 2 & -1 & 0 & \cdots\cdots\cdots\cdots\cdots \\ 0 & -1 & 2 & -1 & \cdots\cdots\cdots\cdots\cdots \\ \cdots & \cdots\cdots\cdots\cdots\cdots\cdots\cdots\cdots\cdots \\ \cdots\cdots\cdots\cdots\cdots & -1 & 2 & -1 & 0 \\ \cdots\cdots\cdots\cdots\cdots & 0 & -1 & 2 & -1 \\ \cdots\cdots\cdots\cdots\cdots & 0 & 0 & -1 & 2 \end{bmatrix}$$

which is a $z \times z$ square matrix. (See footnote on page 167.) This matrix can now be diagonalized by the approach mentioned in the text:

$$[\varphi^{-1}][R][\varphi] = [\Lambda] \tag{c}$$

Operation by $[\varphi^{-1}]$ from the left on equation (b) gives

$$[\varphi^{-1}][C][\dot{X}] = -B[\varphi^{-1}][R][\varphi][\varphi^{-1}][C][X] \tag{d}$$

In equation (d) we took advantage of the fact that $[\varphi^{-1}][\varphi] = [I]$, given in equation (7-49). Now define normal coordinates

$$[q] = [\varphi^{-1}][C][X] \tag{e}$$

and

$$[\dot{q}] = [\varphi^{-1}][C][\dot{X}] \tag{f}$$

which are slightly different from equation (7-51). This is done strictly for mathematical convenience and has no consequence in the final solution. Equation (d) can thus be recast as

$$[\dot{q}] = -B[\Lambda][q] \tag{7-52}$$

where $[q]$ and $[\dot{q}]$ are column matrices and $[\Lambda]$ is a diagonal matrix.

Now diagonalize $[R]$ by rewriting equation (c) as

$$[R][\varphi] = [\varphi][\Lambda] \tag{g}$$

Equation (g) requires that the corresponding elements in the matrices on both sides must be equal after the multiplication has been carried out. Thus, for each value of Λ_{ii}, we obtain a set of linear equations as follows (the reader may wish to verify them for a 3×3 matrix):

$$(2 - \lambda)\varphi_{11} - \varphi_{12} = 0$$
$$-\varphi_{21} + (2 - \lambda)\varphi_{22} - \varphi_{23} = 0$$
$$-\varphi_{32} + (2 - \lambda)\varphi_{33} - \varphi_{34} = 0$$
$$\dotfill \tag{h}$$
$$-\varphi_{z,z-1} + (2 - \lambda)\varphi_{zz} = 0$$

where φ_{ij} is the ijth term in $[\varphi]$.

In order to solve the set of equations given above, note that they all have the same form:

$$-\varphi_{m-1} + (2 - \lambda)\varphi_m - \varphi_{m+1} = 0 \tag{i}$$

with the condition that

$$\varphi_0 = \varphi_{z+1} = 0 \tag{j}$$

Equation (i) can be treated and solved as a difference equation. In operator notation it is just

$$[-E^{-1} + (2 - \lambda) - E]\varphi_m = 0 \tag{k}$$

where the operator E displaces a function in the positive direction and E^{-1} in the negative direction, i.e.,

$$E\varphi_m = \varphi_{m+1}$$
$$E^{-1}\varphi_m = \varphi_{m-1}$$

Now to solve equation (k), assume that the solution has the following form:

$$\varphi_m = \beta e^{m\alpha} \tag{l}$$

where β is a constant and α is a function to be determined. Insertion of equation (l) into equation (k) results in

$$\beta e^{m\alpha}[-e^{-\alpha} + (2 - \lambda) - e^{\alpha}] = 0 \tag{m}$$

In order for equation (m) to yield nontrivial results, it is necessary that

$$-e^{-\alpha} + (2 - \lambda) - e^{\alpha} = 0$$

or

$$\frac{2 - \lambda}{2} = \frac{e^\alpha + e^{-\alpha}}{2} = \cosh \alpha \tag{n}$$

Equation (n) is satisfied by two values of α, namely $+\alpha$ and $-\alpha$, since $\cosh(+\alpha) = \cosh(-\alpha)$. Thus, the solution to the difference equation also must be satisfied by two functions (i.e., $\beta e^{+m\alpha}$ and $\beta e^{-m\alpha}$). The general solution is then just

$$\varphi_m = P_1 \beta e^{m\alpha} + P_2 \beta e^{-m\alpha} \tag{o}$$

where the P's are constants. Since

$$\sinh x = (e^x - e^{-x})/2$$

and

$$\cosh x = (e^x + e^{-x})/2$$

we can have the following equivalent general solution:

$$\varphi_m = M_1 \sinh(m\alpha) + M_2 \cosh(m\alpha) \tag{p}$$

where the M's are constants. For $m = 0$, equation (j), the boundary condition, demands

$$\varphi_0 = 0 = M_2 \tag{q}$$

and

$$\varphi_{z+1} = 0 = \sinh(z + 1) \tag{r}$$

since M_1 is a constant. But the value of the hyperbolic sine is zero if the argument is an integral multiple of $(i\pi)$, where $i = \sqrt{-1}$. Therefore,

$$\alpha = ip\pi/(z + 1), \qquad p = 1, 2, 3, \ldots, z \tag{s}$$

Combination of equations (s) and (n) finally gives the desired result:

$$
\begin{aligned}
\lambda_p &= 2 - 2 \cosh[ip\pi/(z + 1)] \\
&= 2\{1 - \cos[p\pi/(z + 1)]\} \\
&= 4 \sin^2[p\pi/2(z + 1)]
\end{aligned} \tag{7-53}
$$

where $p = 1, 2, 3, \ldots, z$.

Problems

1. Calculate the slope of the curve in Figure 7-3b and its limiting value at times much greater than τ.

2. Develop expressions for the complex modulus and compliance for a Maxwell body
   (a) From the stress relaxation modulus of the body using the phenomenological theory of Chapter 2.
   (b) From the equation of motion of the Maxwell model equation (7-5) assuming a sinusoidal strain application.

3. Develop expressions for D', D'', E', and E'' for a Voigt element.
   (a) From equation (7-25) using the phenomenological theory of Chapter 2.
   (b) From the equation of motion of the Voigt element, equation (7-22) assuming a sinusoidal stress application. (Notice the appearance of a transient term when the boundary condition $\varepsilon(0) = 0$ is used.)

4. Derive an expression for the complex modulus of a Maxwell-Weichert model subjected to a sinusoidal strain. Show that the complex compliance is not obtainable as a simple analytical function.

5. Derive an expression for the complex modulus of z Maxwell elements in series assuming a sinusoidal strain application.

6. Bueche[3] has shown that the flexible bead-spring model leads to the following expression for the creep compliance of an undiluted polymer.

$$D(t) = \frac{8}{3NkT\pi^2} \sum \frac{1}{p^2} [1 - e^{-t/\tau_p}]$$

where $p = 1, 3, 5, \ldots, z$.

In this case the retardation times are given by

$$\tau_p = \frac{\rho a^2 z^2}{3\pi^2 kTp^2} \qquad p = 1, 3, 5, \ldots, z$$

Show
   (a) The retardation times can be rewritten as

$$\tau_p = \frac{12\eta}{N\pi^2 kTp^2}$$

(b) The partial compliances of the equivalent Voigt-Kelvin model are

$$D_p = \frac{8}{3NkT\pi^2 p^2}$$

7. Determine the parameters for a two-element Maxwell-Weichert model $(E_1, E_2, \tau_1, \tau_2)$ which would give a $\log E''(\omega)$ versus $\log \omega$ response similar to that shown below.

8. Show that equation (3-7) yields the viscosity for a Maxwell-Weichert model.

9. Calculate the slope of a stress relaxation master curve at some time t when $\tau_{min} < t < \tau_{max}$ where

$$\tau_p = \frac{\tau_{max}}{p}, \quad \text{and} \quad E_p = k$$

References

1. P. E. Rouse, *J. Chem. Phys.*, **21**, 1272 (1953).

2. W. L. Peticolas, *Rubber Chem. Tech.*, **36**, 1422 (1963).

3. F. Bueche, *J. Chem. Phys.*, **22**, 603 (1954).

4. B. H. Zimm, *J. Chem. Phys.*, **24**, 269 (1956).

5. P. E. Rouse and K. Sittell, *J. Applied Phys.*, **24**, 690 (1954).

6. J. D. Ferry, *Viscoelastic Properties of Polymers*, 2nd ed., Wiley, New York, 1970, p. 213.

7. R. B. DeMallie Jr., M. H. Birnboim, J. E. Frederick, N. W. Tschoegl, and J. D. Ferry, *J. Phys. Chem.*, **66**, 586 (1962).

8. J. D. Ferry, R. F. Landel, and M. L. Williams, *J. Applied Phys.*, **26**, 359 (1955).

9. M. L. Williams, *J. Polymer Sci.*, **62**, 57 (1962).

10. A. V. Tobolsky and D. B. DuPre, *Adv. Polymer Sci.*, **6**, 103 (1969).

11. A. V. Tobolsky, *J. Chem. Phys.*, **37**, 1575 (1962).

12. J. J. Aklonis, Ph.D. thesis, Princeton University, 1965.

13. M. Shen, W. F. Hall, and R. E. DeWames, *J. Macromol. Sci.*, **C2**, 183 (1968).

14. W. V. Houston, *Principles of Mathematical Physics*, 2nd ed., McGraw-Hill, New York, 1948.

15. L. A. Pipes, *Applied Mathematics for Engineers and Physicists*, 2nd ed., McGraw-Hill, New York, 1958.

16. B. Gross, *Mathematical Structure of the Theories of Viscoelasticity*, Hermann, Paris, 1953.

8 Chemical
Stress
Relaxation

We have treated the behavior of polymers whose chemical nature remains constant throughout the time of experimentation. In these cases, the time dependent viscoelastic properties were ascribed to the molecular dynamics inherent in the given macromolecules. Now, however, we shall turn our attention to another type of time dependent behavior which results from changes in the actual molecular composition of the sample. Such phenomena are in the area of chemical stress relaxation.

A. Definition of the Method

The equilibrium stress experienced by a crosslinked rubber network is given by the equation of state for rubber elasticity, developed in Chapter 6 (equation (6-55)).

$$\sigma = N_0 RT \left(\frac{\overline{r_0^2}}{r_f^2} \right) \left[\lambda - \frac{1}{\lambda^2} \right] \tag{8-1}$$

A stress relaxation experiment carried out on a crosslinked rubber will thus result in a constant stress, σ, which will be proportional to the concentration of network chains. Inasmuch as there is no independent means of estimating the "front factor," $(\overline{r_0^2}/\overline{r_f^2})$, and in view of the other approximations inherent in equation (8-1), we shall take the front factor to be equal to unity in the ensuing discussion. Thus we may rewrite equation (8-1) as

$$\sigma = N_0 RT\left[\lambda - \frac{1}{\lambda^2}\right] \tag{8-2}$$

Consider now a crosslinked rubber which is undergoing a chemical reaction resulting in chain scission. If the material is subjected to stress relaxation during this process, it is clear that an equilibrium value of stress will not be attained since the concentration of network chains is not constant. We may rewrite equation (8-2) for this situation as

$$\sigma(t) = N(t)RT\left[\lambda - \frac{1}{\lambda^2}\right]. \tag{8-3}$$

where now both the stress and the concentration of network chains are functions of time. In the limit of long times, the final result of the stress relaxation experiment performed on a degrading network will be the decay of stress to zero, just as in the case of a linear amorphous polymer above T_g (Chapter 3). At the beginning of the stress relaxation experiment, equation (8-3) becomes

$$\sigma(0) = N(0)RT\left[\lambda - \frac{1}{\lambda^2}\right] \tag{8-4}$$

Dividing equation (8-3) by equation (8-4) results in

$$\frac{\sigma(t)}{\sigma(0)} = \frac{N(t)}{N(0)} \tag{8-5}$$

Equation (8-5) provides a direct measure of the extent of degradation of a crosslinked rubber and forms the basis of the chemical stress relaxation technique. That is, measurement of the stress at constant temperature and strain in a degrading rubber network provides a measure of the concentration of network chains as a function of time through equation (8-5).

B. The Two-Network Theory

There are two major limitations to the method presented above. The first of these lies in the use of equation (8-2). It was made clear in the development of Chapter 6 that equation (8-2) is only a first approximation to the true behavior of the system and chemical stress relaxation thus is subject to all the approximations involved in the simple rubber elasticity theory. The second limitation is somewhat less obvious and depends on the details of the degradation process occurring in the network under investigation. The central point is that $N(t)$ in equation (8-5) refers to all the load bearing network chains. If the degradation process involves additional crosslinking as well as chain scission, it is possible that some of the new network chains formed during degradation may be load bearing and some may not. If this is the case, equation (8-5) will no longer provide a measure of the total concentration of network chains at any time and some additional assumption must be introduced in order to deal with the situation quantitatively. It is therefore generally assumed that any new network chains formed during degradation are nonload bearing members of the network and thus do not contribute to $N(t)$ in equation (8-5). This is known as the "two-network" hypothesis and was first introduced by Andrews, Tobolsky, and Hanson.[1] The validity of the two-network hypothesis has been the subject of several investigations[2,3] and the question of its general applicability remains open. Because of its simplicity and success in interpreting data in many systems, we shall assume it to be correct in the discussion which follows unless specifically noted otherwise.

The application of the two-network hypothesis to experimental data involves two types of relaxation measurements, continuous and intermittent. The continuous measurement is the "normal" stress relaxation experiment already discussed in which the polymer is stretched to a fixed length as quickly as possible and maintained at that length and constant temperature while the stress is followed as a function of time. In such a measurement, $N(t)$ can be calculated from equation (8-5). However, $N(t)$ obtained in this way refers only to the load bearing chains in the network and, in the light of the two-network theory, any new chains formed as a result of degradation will contribute nothing to the stress. As a result, $N(t)$ is that portion of the network which has not undergone chain scission at time t. In addition, any chains which have undergone scission at time t and are then reformed cannot contribute to $N(t)$. In the intermittent measurement the sample is stretched to a fixed length as quickly as possible, the stress is measured, and the sample is returned to its original, unstrained condition. This process is repeated at time intervals such that the sample remains unstrained for much longer periods than it is strained. In such a situation, we may again calculate $N(t)$ from equation (8-5). Now, however, $N(t)$ refers to both the portion of the original network remaining at time t and the new network formed by additional crosslinking during degradation. This means that it is possible to separate the degradative process into its component parts of crosslinking reactions and scission reactions. In the case where degradation involves only chain scission and chains once cut are never reformed, both intermittent and

Figure 8-1. $\sigma(t)/\sigma(0)$ Versus Time for the Initiated Oxidation of Hydrocarbon Rubbers. Continuous and Intermittent Measurements Fall on the Same Curve.

continuous measurements yield identical values for $N(t)$. An example is the initiated oxidation of hydrocarbon rubbers illustrated in Figure 8-1.

In order to interpret the kinetics of the degradation process fully, it is necessary to be able to calculate explicitly the number of scission events occurring in the network as a function of time. Unfortunately, this cannot be done in general without introducing assumptions in addition to those used in arriving at equation (8-5) and the two-network hypothesis. In spite of this, a number of relationships have been derived which have proven very useful in the interpretation of chemical stress relaxation experiments. In all cases, the starting point is equation (8-5). Some of the more important expressions will be developed in the following discussion.

C. Irreversible Chain Scission

The first case to be considered is that in which the degradation process consists of the random scission of network chains. In addition, it is postulated that a chain, once cut, is never reformed. In such a case of irreversible chain scission, the intermittent and continuous measurements yield identical results as shown in Figure 8-1 and there is no need to invoke the two-network hypothesis. The most widely applied relationships for computing the time dependence of chain scission under these circumstances are those of Tobolsky[4] and Berry and Watson.[5] The assumptions involved in the Tobolsky equation are that all network chains are elastically active and of the same length, i.e., the crosslinks are uniformly distributed. Let q be the number of scission events per unit volume that have occurred at time t; $M(0)$ be the concentration of chain segments present initially; and x be the number of segments per chain. (These segments refer to the individual entities in the chain which are capable of undergoing scission. In a hydrocarbon type rubber the segments are the backbone carbon-carbon links.) The probability that a particular chain segment has been cut at time t must be $q/M(0)$ since q cuts per unit volume have occurred at time t. Also, since a chain segment has either been cut at time t or it has not been cut, the probability that a given segment has not been cut must be $1 - q/M(0)$. Since all the chains have x segments, the fraction of chains per unit volume remaining uncut at time t is $[1 - q/M(0)]^x$. According to the above, we may write

$$N(t) = N(0) \, [1 - q/M(0)]^x \qquad (8\text{-}6)$$

Also

$$N(0) = M(0)/x \qquad (8\text{-}7)$$

Therefore

$$N(t)/N(0) = \sigma(t)/\sigma(0) = [1 - q/N(0)x]^x \qquad (8\text{-}8)$$

For x large, which is always the case within the range of applicability of the kinetic theory of rubber elasticity, the right hand side of equation (8-8) can be approximated by $e^{-q/N(0)}$ (Problem 8-2).

Substitution into equation (8-8) yields

$$\sigma(t)/\sigma(0) = \exp[-q/N(0)] \qquad (8\text{-}9)$$

or

$$q = -N(0) \ln \sigma(t)/\sigma(0) \qquad (8\text{-}10)$$

Equation (8-10), the Tobolsky equation, has been extensively applied, particularly to the important problem of the oxidative degradation of hydrocarbon networks.

Berry and Watson attempted to refine equation (8-10) by removing the uniform chain length assumption on the grounds that it is unrealistic to expect the distribution of crosslinks to be uniform in real networks. To do this they assumed that the network chains have a range of sizes characterized by the so-called "random" or "most probable" distribution. This is identical to the distribution of molecular weights which results from most free-radical and condensation polymerization processes. Familiarity with molecular weight distributions is assumed here. In this case the concentration of network chains containing x segments becomes

$$N(x) = N(1 - p)^{x-1}p \qquad (8\text{-}11)$$

where N is the total concentration of network chains of all sizes and p is the size parameter (the extent of reaction in the molecular weight distribution case; in the network situation, $p = 1/\bar{x}$).

The initial conditions become

$$N[x,0] = N(0) \, (1 - p)^{x-1}p \qquad (8\text{-}12)$$

In analogy with the derivation of equation (8-10), the fraction of uncut chains of size x per unit volume at time t is $[1 - q/M(0)]^x$. Thus

$$N[x,t] = N[x,0] (1 - q/M(0))^x \qquad (8\text{-}13)$$

Substituting equation (8-12) into equation (8-13) results in

$$\frac{N[x,t]}{N(0)} = p(1 - p)^{x-1}[1 - q/M(0)]^x \qquad (8\text{-}14)$$

In addition, it is clear that

$$\sum_{x=1}^{\infty} N[x,t] = N(t) \qquad (8\text{-}15)$$

Therefore

$$\sum_{x=1}^{\infty} N(x,t)/N(0) = N(t)/N(0) = \sigma(t)/\sigma(0)$$

$$= \sum_{x=1}^{\infty} p(1 - p)^{x-1}[1 - q/M(0)]^x \qquad (8\text{-}16)$$

The right hand side of equation (8-16) is a geometric progression which sums to

$$\sum_{x=1}^{\infty} p(1 - p)^{x-1}[1 - q/M(0)]^x = \frac{p[1 - q/M(0)]}{1 - (1 - p)[1 - q/M(0)]} \qquad (8\text{-}17)$$

Now, since $M(0) = N(0)/p$, we have

$$\frac{p[1 - q/M(0)]}{1 - (1 - p)[1 - q/M(0)]} = \frac{pN(0) - qp^2}{pN(0) - qp^2 + qp} \qquad (8\text{-}18)$$

Since $p < 1$, terms in p^2 may be neglected in comparison to terms in p, leading to the result

$$\sigma(t)/\sigma(0) = N(0)/[M(0) + q] \qquad (8\text{-}19)$$

or

$$q = N(0) [\sigma(0)/\sigma(t) - 1] \qquad (8\text{-}20)$$

Equation (8-20) suffers from the same limitations as equation (8-10) except that the restriction of uniform chain length has been removed. The question arises whether the random distribution of chain sizes in a rubber network is more reasonable than uniform chain sizes. The random distribution emphasizes very short chains, and, because of steric requirements, it would not be expected that a real network would contain such a preponderance of very short chains. In addition, such short chains, if present, would not obey the Gaussian statistics on which both equation (8-10) and equation (8-20) are fundamentally based. Considerations of this sort lead to the conclusion that equations (8-10) and (8-20) are limiting cases and that the true situation must lie somewhere between these two extremes. From the practical point of view, a decision must be made as to which relationship should be applied to the system under consideration. It can easily be shown (Problem 3) that both equation (8-10) and equation (8-20) give essentially the same results for q in the early stages of the scission process so that the choice becomes largely a matter of personal preference provided measurements are restricted to the initial phase of the degradation reaction. Figure 8-2 is a plot of q versus time for both equations (8-10) and (8-20) in which the above point is illustrated.

Figure 8-2. Comparison of the Time Dependence of q as Calculated from Equation (8-10) and Equation (8-20).

D. Reversible Chain Scission

Having treated the situation in which chains once cut are never reformed, we now must establish relationships for reversible scissions in which the cut chains regenerate.[6,7] It will be remembered that in the simplest case of irreversible scission treated already, it is unnecessary to invoke the two-network hypothesis. Any variations of irreversible degradation such as additional crosslinking reactions are still interpretable on the basis of equation (8-10) or (8-20) with the application of the two-network hypothesis. In order to treat the reversible scission case, the two-network hypothesis is necessary even in the simplest situation. In the continuous experiment, only the first scission which occurs in a particular chain is important. No matter how many times the chain is cut, only the first cut will contribute to stress relaxation. Since the chains regenerate, the total concentration of network chains remains at $N(0)$ throughout the experiment and the intermittent value of $\sigma(t)/\sigma(0)$ will thus also remain unity. In the development of the reversible scission equations it is again convenient to treat the extreme case of a uniform network on the one hand and a "random" network on the other.

For the uniform network the probability that a given chain has never been cut at time t becomes $[1 - 1/N(0)]^q$. Now, by the two-network hypothesis, only the chains which have never been cut at time t will be active in supporting the stress. That is, $N(t)$ is the concentration of network chains which have never been cut at time t. Therefore

$$N(t) = N(0) [1 - 1/N(0)]^q \qquad (8\text{-}21)$$

and

$$N(t)/N(0) = \sigma(t)/\sigma(0) = [1 - 1/N(0)]^q \qquad (8\text{-}22)$$

Expanding the right hand side of equation (8-22) using the binomial theorem, there results

$$\sigma(t)/\sigma(0) = 1 - q/N(0) = \exp[-q/N(0)] \qquad (8\text{-}23)$$

or

$$q = -N(0) \ln [\sigma(t)/\sigma(0)] \qquad (8\text{-}24)$$

The result is identical to equation (8-10) obtained for the irreversible scission of a uniform network.

For reversible scission in a random network, the situation is somewhat more complex. In this case the number of cuts per unit volume in chains x segments in length at time t is given by $[qxN(x,0)/\sum xN(x,0)]$. It immediately follows from this that the probability that a particular chain x segments in length has never been cut at time t is

$$\left[1 - \frac{1}{N(x,0)}\right]^{\frac{qxN(x,0)}{\Sigma xN(x,0)}}$$

Following the same arguments used in the uniform network case, the concentration of uncut chains of x segments at time t is

$$N(x,t) = N(x,0)\left[1 - \frac{1}{N(x,0)}\right]^{\frac{qxN(x,0)}{\Sigma xN(x,0)}} \tag{8-25}$$

In the light of equations (8-11) and (8-14), equation (8-25) may be modified to

$$\frac{\sum N(x,t)}{N(0)} = N(t)/N(0) = \sum_{x=1}^{\infty}(1-p)^{x-1}p\left[1 - \frac{1}{N(x,0)}\right]^{\frac{qxN(x,0)}{\Sigma xN(x,0)}} \tag{8-26}$$

Proceeding in a fashion analogous to that used in deriving equations (8-10), (8-20), and (8-24), the expression

$$\left[1 - \frac{1}{N(x,0)}\right]^{\frac{qxN(x,0)}{\Sigma xN(x,0)}}$$

is expanded by the binomial theorem

$$[1 - 1/N(x,0)]^{\frac{qxN(x,0)}{\Sigma xN(x,0)}} = 1 - \frac{qx}{\sum xN(x,0)}$$

$$+ \cdots \cong e^{-\frac{qx}{\Sigma xN(x,0)}} \tag{8-27}$$

Using equation (8-11) it is clear that

$$\sum xN(x,0) = N(0)p \sum x(1 - p)^{x-1} = \frac{N(0)}{p} \qquad (8\text{-}28)$$

Since $p = 1/x$; $N(0)/p = M(0)$ and equation (8-27) may be rewritten as

$$[1 - 1/N(x,0)]^{\frac{qxN(x,0)}{\Sigma xN(x,0)}} = \exp[-qx/M(0)] \qquad (8\text{-}29)$$

In addition, equation (8-26) becomes

$$N(t)/N(0) = \sigma(t)/\sigma(0) = \sum_{x=1}^{\infty} (1 - p)^{x-1} pe^{-\frac{qx}{M(0)}} \qquad (8\text{-}30)$$

The summation on the right hand side of equation (8-30) is easily evaluated since it is simply a geometric series with the first term being $p \exp[-q/M(0)]$ and the common ratio being

$$[(1 - p) \exp(-q/M(0))]$$

Thus

$$\sigma(t)/\sigma(0) = \frac{p \exp[-q/M(0)]}{1 - (1 - p) \exp[-q/M(0)]} \qquad (8\text{-}31)$$

which yields, upon rearrangement

$$q/M(0) = \ln \left[p\left(\frac{\sigma(0)}{\sigma(t)} - 1\right) + 1 \right] \qquad (8\text{-}32)$$

E. Chemical Permanent Set

The central role of the two-network hypothesis in chemical stress relaxation has already been dealt with at some length. An experimental test of this hypothesis is possible through the phenomenon of chemical permanent set. In general, if a sample of a degrading rubber network is strained, it will not return to its original length when released but rather will assume a new equilibrium length intermediate between its original length and its strained length. In the limit of long times when the stress has decayed to zero, the new equilibrium length will be identical to the strained length. This is the phenomenon of chemical permanent set. For convenience, we shall express it as the percentage of the deformation retained by the sample at any time. That is,

$$\text{p.s.} = \frac{L_s - L_u}{L_x - L_u} \times 100 \tag{8-33}$$

where L_s is the new equilibrium unstrained length of the sample at time t, L_u is the original unstrained length and L_x is the strained length. In terms of the two-network hypothesis, the permanent set length may be thought of as that length which represents a balance of two opposing forces: the restoring force exerted by the network in equilibrium with the original length of the sample, and the extending force exerted by the network in equilibrium with the strained length of the sample. Using an alternate form of the equation of state for rubber elasticity derived in the Appendix at the end of this chapter, we have

$$\sigma_u' = N_u kT\left[\left(\frac{L_s}{L_u}\right)^2 - \left(\frac{L_u}{L_s}\right)\right] \tag{8-34}$$

$$\sigma_x' = N_x kT\left[\left(\frac{L_s}{L_x}\right)^2 - \left(\frac{L_x}{L_s}\right)\right] \tag{8-35}$$

where σ_u' is the true stress due to the concentration of network chains N_u which are at equilibrium at the original length of the sample, and σ_x' is the true stress due to the concentration of network chains N_x which are in equilibrium with the strained length of the sample. At the permanent set length of the sample

$$\sigma_u' = -\sigma_x \tag{8-36}$$

From equation (8-34) and equation (8-35)

$$\frac{N_u}{N_x}\left[\left(\frac{L_s}{L_u}\right)^2 - \frac{L_u}{L_s}\right] = \frac{L_x}{L_s} - \left(\frac{L_s}{L_x}\right)^2 \tag{8-37}$$

Substitution of equation (8-33) into equation (8-37) yields an expression for the permanent set on the basis of the two-network theory

$$\text{p.s.} = C_3\left[\left(\frac{C_1}{C_2(N_u/N_x) + 1} + 1\right)^{\frac{1}{3}} - 1\right] \tag{8-38}$$

where $C_1 = (L_x/L_u)^3 - 1$; $C_2 = (L_x/L_u)^2$; $C_3 = 100/[(L_x/L_u) - 1]$.

All the quantities in equation (8-38) are experimentally accessible. N_u/N_x is obtainable from the expression

$$\frac{N_u}{N_x} = \frac{\sigma'(t)/\sigma'(0)_{\text{continuous}}}{\sigma'(t)/\sigma'(0)_{\text{intermittent}} - \sigma'(t)/\sigma'(0)_{\text{continuous}}} \quad (8\text{-}39)$$

Equation (8-39) follows from the fact that the continuous measurement is a measure of the concentration of network chains in the sample which have not undergone scission at time t while the intermittent measurement reflects the total concentration of network chains in the sample including new chains formed as a result of degradation. It is thus possible to calculate the chemical permanent set on the basis of the two network theory using equations (8-38) and (8-39) and to compare it with the experimental results. Figure 8-3 makes this comparison for oxidizing hydrocarbon rubber networks

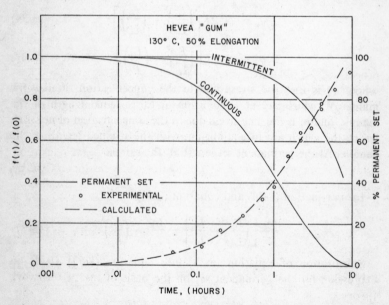

Figure 8-3. Comparison Between Calculated and Observed Chemical Permanent Set for Oxidizing Hydrocarbon Rubber Networks. After R. D. Andrews et al., *J. Appl. Phys.*, **17**, 352 (1946) with permission of the American Institute of Physics and the Authors.

and the agreement is seen to be excellent. It is obviously very useful to carry out such a test of the two network hypothesis when undertaking chemical stress relaxation studies of novel systems.

The necessary machinery has now been developed to examine applications of chemical stress relaxation to specific degrading systems. It is beyond the scope of this treatment to present a detailed account of the chemical stress relaxation method to all the systems which have been studied.

Rather, a brief summary is presented in what follows and the reader is referred to the original literature for more detailed information. Chemical stress relaxation has been used to study networks degrading under the influence of corrosive gases, heat, and radiation of various kinds. Both reversible and irreversible scission processes have been examined.

F. Oxidative Degradation

An outstanding example of an irreversible scission process is the oxidative degradation of hydrocarbon rubber networks. This process is extremely complex and has been studied by many workers using a variety of techniques. Aside from the intrinsic interest of the problem, its obvious technological importance has led to the very numerous and extensive investigations which have been carried out. Some of the results of chemical stress relaxation studies on the initiated oxidation of hydrocarbon elastomers will be summarized. (The problem of uninitiated oxidation, or autoxidation, is very much more complex and will not be discussed here. For a review of autoxidation studies, see reference 8.)

The general kinetic scheme for the initiated oxidation of hydrocarbon rubbers is:

Initiation: Initiator $\xrightarrow{k_d}$ 2R'·

$$R'· + RH \xrightarrow{k_1} R· + R'H$$

Propagation: $R· + O_2 \xrightarrow{k_2} RO_2·$

$$RO_2· + RH \xrightarrow{k_3} ROOH + R·$$

Termination: $2R\cdot \xrightarrow{k_4} R–R$ (recombination)

$RO_2\cdot + R\cdot \xrightarrow{k_5} ROOR$

$2RO_2\cdot \xrightarrow{k_6}$ nonradical products

In the above scheme, $R'\cdot$ is a free radical formed by the decomposition of the initiator and $R\cdot$ is a free radical polymer chain.

From the scheme outlined above, it is not possible to identify the kinetic step in which chain scission takes place. However, it is precisely the scission reaction which leads to chemical stress relaxation. Thus, the method provides a convenient means of determining the kinetic step during which the scission reaction takes place. As an illustration may be cited the work of Norling, Lee, and Tobolsky.[9] These authors studied the benzoyl peroxide initiated oxidation of elastomers containing carbon-carbon crosslinks. The Tobolsky equation, equation (8-10), was used to calculate q. From a plot of q as a function of time the initial rate of scission was determined. Such measurements were carried out with a series of initiator concentrations. The results showed that the initial rate of scission has a linear dependence on initiator concentration (Figure 8-4). Stated mathematically

$$\left(\frac{dq}{dt}\right)_0 = e_s(2k_dI_0) + \left(\frac{dq}{dt}\right)_{th} \qquad (8\text{-}40)$$

where $(dq/dt)_0$ is the initial rate of scission, e_s is the efficiency of scission and reflects the fact that not every radical formed by the decomposition of the benzoyl peroxide is effective in starting oxidation, k_d is the rate constant for the decomposition of the initiator, I_0 is the initiator concentration, and $(dq/dt)_{th}$ is the rate of scission in the absence of the initiator. k_d is known for benzoyl peroxide from other kinetic investigations so that the slope of a plot of $(dq/dt)_0$ versus I_0 yields values of e_s. Values of e_s for some elastomers obtained in this way are presented in Table 8-1. Intermittent relaxation measurements on these substances indicated that no additional crosslinking took place during the degradation process. This result, coupled with the fact that the scission process is irreversible, showed that there was no need to invoke the two network hypothesis in interpreting the data. It was also found that e_s is independent of

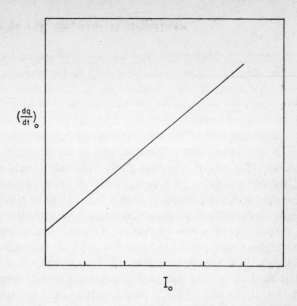

$\left(\dfrac{dq}{dt}\right)_0$

I_0

Figure 8-4. Dependence of Initial Rate of Scission on Initiator Concentration for Initiated Oxidation of Hydrocarbon Rubber Networks.

TABLE 8-1. Efficiency of Scission Values, Benzoyl Peroxide Initiates 60–100°C [After Norling, Lee, and Tobolsky[6]]

Polymer	e_s
Ethylene–propylene copolymer	0.65
Ethylene–propylene terpolymer	0.60
Amorphous polypropylene	0.60
Polyvinylmethyl ether	0.4
Polypropylene oxide	0.65
Styrene–butadiene coploymer	0.65
Natural rubber	0.6
Polymethylacrylate	0.2
Polyethylacrylate	0.16
Polypropylacrylate	0.07
Polybutylacrylate	0.03

kinetic chain length. This means that scission does not occur in the propagation step, for, if scission did take place during propagation, it is apparent that e_s should be a function of the kinetic chain length. It remains to assign the scission process to either the initiation or termination step. In order to do this, further measurements are necessary since the rate of initiation is the same as the rate of termination at the steady state. Since oxygen is not involved in the initiation step, the rate of chemical stress relaxation should not be affected by the absence of oxygen if scission takes place in the initiation step. Relaxation measurements carried out in a nitrogen atmosphere revealed that the relaxation rate was very much reduced, leading to the conclusion that scission must occur in the termination step. The highest values of e_s in Table 8-1 are 0.60–0.65 which are approximately those for the initiation efficiency of benzoyl peroxide when it is used to initiate free radical polymerizations. Inspection of Table 8-1 shows that the e_s's of the poly-n-alkyl acrylates are anomalously low. This is apparently associated with side chain oxidation which would not, of course, contribute to the relaxation process. The efficiencies are seen to decrease with increasing side chain length as is to be expected on the basis of this mechanism.

G.   Interchange Networks: The Polysulfides

As an example of chemical stress relaxation involving reversible scission may be cited the case of the polysulfides or "Thiokol" elastomers.[10] These polymers have the general structure

where R is an alkyl group and S_x is a polysulfide group. Usually x varies from one to five and it is referred to as the sulfur rank of the polymer. The preparation of the polysulfide elastomers is usually done by reacting an alkyl dihalide with an aqueous solution of sodium polysulfide. Crosslinking is introduced by adding the desired quantity of trihalide to the reaction mixture. As a result of this procedure it is possible to vary x, the nature of R, and the degree of crosslinking. It should be pointed out that aqueous sodium polysulfide solutions always contain a distribution of sizes of polysulfide anions so that it is never possible to obtain a polymer with uniform polysulfide repeat units by this technique. The quoted sulfur rank of the polymer is merely an average value.

The outstanding chemical property of the polysulfide elastomers is their ability to undergo interchange reactions by the scission and recombination of a sulfur-sulfur bond.

$$RS_4R + R'S_4R' \rightleftharpoons 2RS_4R'$$
$$RS_2R + R'S_2R' \underset{}{\overset{cat.}{\rightleftharpoons}} 2RS_2R'$$

These reactions are responsible for the chemical stress relaxation process in the polysulfides. It may be seen that the scission is reversible and this is confirmed experimentally by a study of the intermittent relaxation which remains constant at a value of $\sigma(t)/\sigma(0) = 1$ (Figure 8-5). Chemical stress relaxation has proven useful in the elucidation of the kinetics of scission in these polymers.[10] The experimental results of such studies may be summarized as follows. First, for a polyethylene tetrasulfide series of varying degrees of crosslinking, the relative stress versus time curves show simple exponential behavior at all temperatures and all degrees of crosslinking. Stated mathematically

$$\sigma(t)/\sigma(0) = e^{-k't} \tag{8-41}$$

From plots of $\ln \sigma(t)/\sigma(0)$ versus time at different temperatures it is possible to obtain k' as a function of temperature. The activation energy for the stress decay process results from a plot of $\log k'$ versus the reciprocal of the absolute temperature. The results of this analysis show that the activation energy remains constant at a value of 25–26 kcal mole^{-1} through a crosslinking range from 10 mole per cent to 50 mole per cent. Having defined the effect of crosslinking, the next variable to be investigated is the sulfur rank. Over a range of sulfur rank from 1.5 to 4.0 it was found that the stress decay is well represented by equation (8-41) but the value of k' increases with increasing sulfur rank, i.e., the rate of stress relaxation

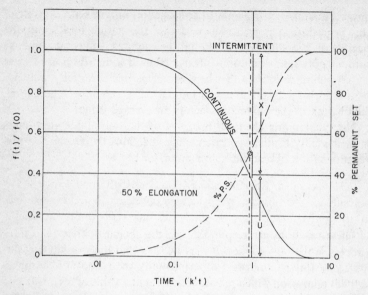

Figure 8-5. Intermittent Relaxation Versus Time for a Polysulfide Rubber Network. After A. V. Tobolsky, *Properties and Structure of Polymers*, Figure 5-20 with the permission of John Wiley & Sons, Inc.

increases with increasing sulfur rank. The activation energies remain constant at the values of 25–26 kcal mole⁻¹ just as in the case of the varying degrees of crosslinking. Finally, a study was conducted on samples containing elemental sulfur physically mixed with the polymers. This was accomplished by immersing the crosslinked samples in molten sulfur where they were observed to swell and to absorb large quantities of sulfur. When the samples were removed from the sulfur bath, the admixed sulfur showed no tendency to crystallize for long periods of time, remaining in a supercooled liquid condition. The experimental results on these samples again show that the relaxation process obeys equation (8-41) and that the activation energy for stress decay is again 25–26 kcal mole⁻¹. However, the elemental liquid sulfur greatly accelerates the relaxation process in the case of polymers of rank two while leaving the relaxation rate essentially unchanged in the case of polymers of higher rank (rank four). In order to formulate a kinetic scheme for stress decay in the crosslinked polysulfides, the following experimental findings are of significance in addition to the results described above.

1. Eight-membered sulfur rings, long chain polymeric sulfur, and low molecular weight alkyl tetrasulfides all undergo thermal homolytic scission with an activation energy of approximately 35 kcal mole^{-1}.

2. Electron spin resonance and magnetic susceptibility data indicate the existence of free radical chain ends in polymeric sulfur at temperatures of around 300°C.

3. Alkyl disulfides are much more stable thermally than alkyl polysulfides. As noted above, the rate of chemical stress relaxation in the polysulfide polymers increases markedly with increasing sulfur rank.

4. Stress relaxation of polysulfide polymers is accelerated by near ultraviolet light and the activation energy for the process is thereby greatly lowered.

Utilizing all the foregoing experimental findings, the following kinetic scheme is proposed for the chemical stress relaxation of the polysulfides.

Initiation: $\quad \sim S_4$ (strained) $\xrightarrow{k_i} \sim S_2\cdot + \cdot S_2\sim$ (unstrained)

or, in the case of admixed sulfur

$$S_8 \text{ (ring)} \xrightarrow{k_i} S_8 \text{ (diradical chain)}$$

Propagation:

$$\sim S_2\cdot + \sim S_x\sim \text{ (strained)} \xrightarrow{k_p} \sim S_x\sim \text{ (unstrained)} + \sim S_2\cdot$$

Termination: $\quad \sim S_2\cdot + \cdot S_2\sim \xrightarrow{k_t} \sim S_4\sim$ (unstrained)

All of the reactions written above are, of course, reversible, but the reverse reactions contribute nothing to $N(t)$ by the two-network hypothesis. It remains to establish that the chain reaction presented schematically above, does indeed satisfy the experimental results. The rate of stress decay is given by

$$\frac{-dN(t)}{dt} = k_i[\sim S_4\sim] + k_p[\sim S_x\sim][\sim S_2\cdot] \qquad (8\text{-}42)$$

The assumption is now made that the observed relaxation takes place mainly in the propagation step. This means that a single scission event in the initiation step produces many scissions in the propagation step. If this is so, the first term of equation (8-42) is small compared to the second and equation (8-42) becomes

$$\frac{-dN(t)}{dt} = k_p[\sim S_x\sim][\sim S_2\cdot] \tag{8-43}$$

At the stationary state, the rate of initiation is equal to the rate of termination,

$$k_i[\sim S_4\sim] = k_t[\sim S_2\cdot]^2 \tag{8-44}$$

or

$$[\sim S_2\cdot] = \left(\frac{k_i}{k_t}\right)^{1/2}[\sim S_4\sim]^{1/2} \tag{8-45}$$

and

$$\frac{-dN(t)}{dt} = k_p\left(\frac{k_i}{k_t}\right)^{1/2}[\sim S_4\sim]^{1/2}[\sim S_x\sim] \tag{8-46}$$

Since it is assumed that stress relaxation occurs in the propagation step, $[\sim S_x\sim]$ can be replaced by $mN(t)$, where m is the number of polysulfide links (linkages susceptible to cleavage) per network chain. This is obviously not strictly correct since scission of an S_4 chain in the initiation step also leads to relaxation of stress. However, if the kinetic chain length is great, the error introduced by this assumption will be small. Also, under these circumstances, $[\sim S_4\sim]$ will remain approximately constant throughout the reaction. Designating $[\sim S_4\sim]$ by I there results

$$\frac{-dN(t)}{dt} = k_p\left(\frac{k_i}{k_t}\right)^{1/2}ImN(t) \tag{8-47}$$

It should be noted that equation (8-47) is equally valid when the initiator is elemental sulfur. In the case of sulfur as the initiator, it is only necessary to set I in equation (8-47) equal to the concentration of elemental sulfur added to the polymer.

In order to compare equation (8-47) to experiment, it is necessary to return to equation (8-41). By the fundamental relation of chemical stress relaxation, equation (8-41) becomes

$$\sigma(t)/\sigma(0) = N(t)/N(0) = e^{-k't} \tag{8-48}$$

which, upon differentiation, yields

$$\frac{-dN(t)}{dt} = k'N(t) \tag{8-49}$$

Comparing equation (8-49) to equation (8-47)

$$k' = k_p\left(\frac{k_i}{k_t}\right)^{1/2} I^{1/2}m \tag{8-50}$$

At a given temperature, all of the quantities in equation (8-50) are independent of sulfur rank with the exception of I. I increases with sulfur rank since it is only the higher polysulfides which are capable of undergoing thermal homolytic scission. As a result, k' also increases with increasing sulfur rank, leading to an increase in the overall rate of stress relaxation as is actually observed. In the same fashion, if the concentration of added elemental sulfur is increased, k' increases and the rate of stress relaxation at a given temperature again increases. In addition, as the degree of crosslinking increases, the number of polysulfide groups per network chain decreases and m decreases, thus reducing the rate of stress relaxation, again in accord with experiment.

Finally, it is necessary to discuss the temperature dependence of the rate constant. From equation (8-50) it immediately follows that the observed activation energy, H', is given by

$$H' = H_p + 1/2(H_i - H_t) \tag{8-51}$$

As already stated, H' amounts to 25–26 kcal in all systems studied. In addition, as has been discussed, H_i is 35 kcal mole^{-1}. Since H_t involves a radical recombination reaction, its magnitude is probably quite small and the assumption is thus made that it is equal to zero. On this basis, H_p becomes approximately 8 kcal mole^{-1}. The only direct measurement of the activation energy for the attack of a polysulfide radical on a polysulfide linkage[11] gives an H_p in the range of 3–6 kcal mole^{-1} in fair agreement with the value obtained above.

Appendix

In Chapter 6, the equation of state for rubber elasticity based on the statistical theory was shown to be

$$\sigma = N_0 RT \left[\lambda - \frac{1}{\lambda^2} \right] \tag{8-2}$$

Here σ is the elastic force per unit *unstrained* cross-sectional area ($\sigma = F/A_0$) and is generally called the "nominal stress." This is the definition of stress used throughout most of this book. It is possible, however, to define the stress in terms of the cross-sectional area of the strained rubber:

$$\sigma' = \frac{F}{A} \tag{a}$$

where σ' is the "true stress."

The equation of state will now be rewritten in terms of the true stress. In equation (8-2), the concentration of network chains (N_0) is defined as the number of chains (N) divided by the volume of the unstrained rubber (V_0); and λ is just L/L_0, the ratio of the lengths of the stretched and unstretched rubber. Equation (8-2) can be rewritten as

$$\frac{F}{A_0} = \left(\frac{NRT}{V_0} \right) \left(\frac{L}{L_0} - \frac{L_0^2}{L^2} \right)$$

or, since $V_0 = A_0 L_0$,

$$F = \left(\frac{NRT}{L_0} \right) \left(\frac{L}{L_0} - \frac{L_0^2}{L^2} \right) \tag{b}$$

Now divide equation (b) by the cross-sectional area of the strained sample to obtain the true stress:

$$\sigma' = \frac{F}{A} = \left(\frac{NRT}{AL_0} \right) \left(\frac{L}{L_0} - \frac{L_0^2}{L^2} \right) \tag{c}$$

Multiplying by L/L and rearranging, one gets

$$\sigma' = \left(\frac{NRT}{V} \right) \left(\frac{L}{L_0} \right) \left(\frac{L}{L_0} - \frac{L_0^2}{L^2} \right) \tag{d}$$

However, we know from Chapter 6 that upon extension the volume of the rubber stays nearly constant; thus,

$$\sigma' = N_0 RT \left(\frac{L^2}{L_0{}^2} - \frac{L_0}{L} \right) \qquad \text{(e)}$$

Equation (e) is an alternative expression of the equation of state for rubber elasticity previously given in equation (8-2).

Problems

1. Plot schematically the time dependence of intermittent and continuous relaxation measurements for degradation processes in which

   (a) Both chain scission and additional crosslinking take place

   (b) Only chain scission takes place but chains reform after being cut.

2. Show that $[1 - q/N(0)x]^x \approx \exp[-q/N(0)]$ for large x.

3. Show that equation (8-10) and equation (8-20) are equivalent at low values of q.

4. Verify that the leading terms in the binomial expansion of $[1 - 1/N(0)]^q$ are 1 and $-q/N(0)$.

5. Evaluate $N(0)p \sum_{x=1}^{\infty} x(1 - p)^{x-1}$.

6. Obtain equation (8-38) from equation (8-33) and equation (8-37) using definitions of C_1, C_2, and C_3.

7. Show that $E' = E_p + 1/2(E_i - E_t)$ starting from equation (8-50).

References

1. A. V. Tobolsky, *Properties and Structure of Polymers*, Wiley, New York, 1960.

2. P. J. Flory, *Chem. Rev.*, **35,** 51 (1944).

3. D. K. Thomas, *Polymer*, **7,** 125 (1966).

4. A. V. Tobolsky, D. J. Metz, and R. B. Mesrobian, *J. Am. Chem. Soc.*, **72,** 1942 (1950).

5. J. P. Berry and W. F. Watson, *J. Polymer Sci.*, **18,** 204 (1955).

6. H. Yu, *J. Polymer Sci.*, **B2,** 631 (1964).

7. A. V. Tobolsky, *J. Polymer Sci.*, **B2,** 631 (1964).

8. J. R. Dunn and J. Scanlon, Chapter 18 in *The Chemistry and Physics of Rubberlike Substances*, L. Bateman, ed., MacLaren, London, 1963.

9. P. M. Norling, T. C. P. Lee, and A. V. Tobolsky, *Rubber Chem. and Tech.*, **38,** 1198 (1965).

10. A. V. Tobolsky and W. J. MacKnight, *Polymeric Sulfur and Related Polymers*, Wiley, New York, 1966, Chapter IV.

11. D. K. Gardner and K. Fraenkel, *J. Am. Chem. Soc.*, **78,** 3279 (1956).

Answers to Problems
◆◆◆◆◆◆◆◆◆◆◆◆◆◆◆◆◆◆◆◆◆◆◆◆◆◆◆◆◆◆◆◆

Chapter 2

2-1. The shear strain wanted is given, according to equation (2-5) as

$$\gamma = \frac{\Delta X}{C} = \frac{0.40\,\text{cm}}{2.0\,\text{cm}} = 0.20$$

At $t = 10^{-4}$, the compliance is numerically equal to $10^{-10}\,\text{cm}^2/\text{dyne}$. Thus, making use of equation (2-9),

$$\sigma_{s_0} = \frac{\gamma(t)}{J(t)} = \frac{\gamma(10^{-4})}{J(10^{-4})} = 0.20 \times 10^{10}\,\text{dyne/cm}^2$$

However, since the sample cross sectional area, $A \times B$, is $4\,\text{cm}^2$ the constant force necessary to observe this strain is

$$F_{s_0} = \sigma_{s_0}(A \times B) = 0.8 \times 10^{10}\,\text{dyne}$$

and division by the acceleration due to gravity yields

$$m = \frac{F_{s_0}}{a} = 8.2 \times 10^6\,\text{gm}$$

Thus 8.2×10^6 grams placed on the sample pan in Figure 2-2 would cause the pointer to move down $0.40\,\text{cm}$ in 10^{-4} seconds. Clearly all the apparatus is assumed to have an infinite modulus and inertial effects are ignored.

2-2. Since G' and G'' are given, our problem is to express $\tan \delta$, $|G|^2$, $|J|^2$, J', and J'' in terms of these given parameters. Equation (2-21) gives

$$\tan \delta = G''/G'$$

Next it will be helpful to place the storage and loss components of both the stress and strain on the same set of stress strain vectors.

217

This is shown in Figure 2-8a which is merely a composite of Figure 2-7. The various vectors are labeled in the table. For clarity, it is now desirable to rotate the triangle ABF so that side AB of that triangle is coincident with side AC of triangle AEC. This is shown in Figure 2-8b. The vectors designated in the table do not change in magnitude under this transformation.

To proceed, AE represents σ', CE represents σ'', and AC represents σ. If these three quantities are divided by the magnitude of γ, they become G', G'', and $|G|$ respectively. Clearly then application of the Pythagorean Theorem gives

$$|G|^2 = G'^2 + G''^2$$

Now from Figure 2-8, relationships of congruent triangles give

$$\frac{BF}{AB} = \frac{CE}{AC}$$

In terms of more relevant quantities

$$\frac{\gamma''}{\gamma} = \frac{\sigma''}{\sigma}$$

Arithmetic rearrangement and division by σ gives

$$\frac{\gamma''}{\sigma} = \sigma'' \frac{\gamma}{\sigma^2} \cdot \frac{\gamma}{\gamma} = \frac{\sigma''}{\gamma} \frac{\gamma^2}{\sigma^2}$$

which is equal to

$$J'' = \frac{G''}{|G|^2} = \frac{G''}{G'^2 + G''^2}$$

Using the same technique

$$J' = \frac{G'}{|G|^2} = \frac{G'}{G'^2 + G''^2}$$

Finally, since $|J|$ and $|G|$ are reciprocals,

$$|J|^2 = \frac{1}{|G|^2} = \frac{1}{G'^2 + G''^2}$$

2-3.   Equation (2-28) may be restated as

$$\sigma(t) = \int_{-\infty}^{t} \frac{d\varepsilon(\mu)}{d\mu} E(t-\mu)\, d\mu \tag{a}$$

for a tensile experiment. The strain history, in terms of the variable μ, is

$$-\infty < \mu \le 0 \qquad \frac{d\varepsilon(\mu)}{d\mu} = 0$$

$$0 < \mu \le t_1 \qquad \frac{d\varepsilon(\mu)}{d\mu} = K$$

$$t_1 < \mu \le 2t_1 \qquad \frac{d\varepsilon(\mu)}{d\mu} = -K$$

$$2t_1 < \mu \le t \qquad \frac{d\varepsilon(\mu)}{d\mu} = 0$$

This strain history can be built into equation (a) to yield

$$\sigma(t) = \int_{0}^{t_1} K E e^{-(t-\mu)/\tau}\, d\mu - \int_{t_1}^{2t_1} K E e^{-(t-\mu)/\tau}\, d\mu$$

which can be integrated to give

$$\sigma(t) = K E\tau[2e^{t/\tau} - e^{2t_1/\tau} - 1]$$

2-4.   Work is usually defined as the integral of $f\,dl$. Here we will consider work per unit volume. Thus

$$W = \int \frac{f\,dl}{V} = \int \frac{f}{A}\frac{dl}{l} = \int \sigma\, d\gamma \tag{a}$$

for a shear experiment. Now let

$$\sigma = \sigma_0 \sin \omega t$$

so that in phase strain is

$$\gamma = \gamma_0 \sin \omega t$$

Substitution of these functions into equation (a) yields

$$W = \int_0^{2\pi} \sigma_0 (\sin \omega t) \gamma_0 \, d \sin \omega t$$

which integrates to zero. Thus the work done in going through one cycle of deformation (ωt goes from 0 to 2π) is zero if the stress and strain are in phase.

For out of phase strain

$$\gamma = \gamma_0 \cos \omega t$$

which yields for the integral work

$$W = \int_0^{2\pi} \sigma_0 (\sin \omega t) \gamma_0 (\cos \omega t) \, d \omega t$$

Here integration gives

$$W = \gamma_0 \sigma_0 \pi$$

Clearly work is done when the stress and strain are out of phase.

2-5. Equation (2-49) gives

$$\frac{1}{p^2} = L(G(t)) \, L(J(t)) \tag{2-49}$$

which is a statement of the relationship between $G(t)$ and $J(t)$ in transform space.

The Laplace transform of the $J(t)$ given is

$$L(J(t)) = \frac{A\Gamma(1 + m)}{p^{m+1}}$$

which yields upon substitution into equation (2-49) the result for the transform of $G(t)$

$$L(G(t)) = \frac{1}{A\Gamma(m + 1)p^{1-m}} \frac{\Gamma(1 - m)}{\Gamma(1 - m)}$$

the last multiplication term being added to facilitate return from transform space. Thus

$$G(t) = \frac{t^{-m}}{A\Gamma(m + 1)\Gamma(1 - m)}$$

Rearrangement and use of the expressions given in the problem yields the desired result.

2-6. Problem 2-6 is similar to 2-5.

Chapter 3

3-1(a)   Equation (3-6) states

$$\log a_T = \frac{-C_1(T - T_g)}{C_2 + T - T_g} \tag{3-6}$$

The apparent energy of activation for viscoelastic relaxation may be defined as

$$\Delta H_a = R \frac{d \ln a_T}{d \frac{1}{T}}$$

which by simple calculus is just

$$\Delta H_a = -RT^2(2.303) \frac{d \log a_T}{dT} \tag{a}$$

Using equation (a) on equation (3-6) yields

$$\Delta H_a = \frac{2.303 C_1 C_2 R T^2}{(C_2 + T - T_g)^2} \tag{b}$$

Since we are interested in the activation energy at T_g, equation (b) is just

$$\Delta H_a = 2.303 \left(\frac{C_1}{C_2}\right) R T_g^2$$

Using the universal constants given in Table 3-2, at 200°K this quantity is about 61 kcal mole^{-1}.

(b)   As $T \gg T_g$ equation (b) becomes

$$\Delta H_a = 2.303 R C_1 C_2$$

which is equal to 4.1 kcal mole^{-1}.

3-2. Let all quantities with primes denote the new reference temperature. First

$$\log a_T(T_g + 50) = -\log a'_T(T_r - 50)$$

As a second equation we might write

$$\log a_T(T_g + 50) = \log a_T(T_g + 25) - \log a'_T(T_r - 25)$$

Thus we have two independent linear equations in two unknowns C'_1 and C'_2 which can be solved. In this case $C'_1 = 8.55$ and $C'_2 = 100$.

3-3. This problem is worked out in reference 9 of Chapter 3.

Chapter 4

4-1. Equation (3-16) may be written as

$$f_M = f_g + \alpha_f(T - T_{gM})$$
$$f_\infty = f_g + \alpha_f(T - T_{g\infty})$$

where f_M is the fractional free volume at some temperature and molecular weight. Here we have assumed α_f and f_g are both independent of molecular weight. These relationships may be introduced into the equation given in the problem to yield equation (4-25).

4.4 (a) The weight fraction of chain ends, w_e, is just the weight of the chain ends divided by the total weight of the polymer. If n_i is the number of moles of polymer with molecular weight M_i, w_e is given as

$$w_e = \frac{\text{weight of chain ends}}{\text{total weight}} = \frac{\sum_i n_i M_e}{\sum_i n_i M_i} = M_e\left(\frac{\sum_i n_i}{\sum_i n_i M_i}\right)$$

The term in parentheses, however, is just the reciprocal of the number average molecular weight of the polymer, $\bar{M}_n$, so

$$\bar{M}_n = \frac{M_e}{w_e}$$

Substitution of this relationship into equation (4-25) yields the desired result.

(b) Rewriting the equation given in terms of relevant parameters, remembering that $w_p + w_e = 1$, gives

$$T_g = T_g{}^\infty(1 - w_e) + T_{g_c}w_e + Kw_e - K(w_e)^2$$

Since w_e is small for polymers, the last term can be neglected and

$$T_g = T_g{}^\infty - w_e(T_g{}^\infty - T_{g_e} - K)$$

This expression is identical to the desired result if we identify the term in brackets with c/M_e.

Chapter 5

5-1.

$$\overline{r^2} = \frac{1}{1}\sum_{i=1}^{1}(\mathbf{r}_i)^2 = (\mathbf{r}_1)^2$$

$$\overline{r^2} = \left(\sum_{j=1}^{4}\mathbf{l}_j\right)^2 \tag{5-2}$$

$$\overline{r^2} = \mathbf{l}_1 \cdot \mathbf{l}_1 + \mathbf{l}_1 \cdot \mathbf{l}_2 + \mathbf{l}_1 \cdot \mathbf{l}_3 + \mathbf{l}_1 \cdot \mathbf{l}_4 +$$
$$\mathbf{l}_2 \cdot \mathbf{l}_1 + \mathbf{l}_2 \cdot \mathbf{l}_2 + \mathbf{l}_2 \cdot \mathbf{l}_3 + \mathbf{l}_2 \cdot \mathbf{l}_4 +$$
$$\mathbf{l}_3 \cdot \mathbf{l}_1 + \mathbf{l}_3 \cdot \mathbf{l}_2 + \mathbf{l}_3 \cdot \mathbf{l}_3 + \mathbf{l}_3 \cdot \mathbf{l}_4 +$$
$$\mathbf{l}_4 \cdot \mathbf{l}_1 + \mathbf{l}_4 \cdot \mathbf{l}_2 + \mathbf{l}_4 \cdot \mathbf{l}_3 + \mathbf{l}_4 \cdot \mathbf{l}_4$$

5.3. Equation (5-32) may be written as

$$\overline{r^2} = \frac{\displaystyle\int_0^\infty r^4 e^{-b^2 r^2}\,dr}{\displaystyle\int_0^\infty r^2 e^{-b^2 r^2}\,dr}$$

after using equation (5-30) to eliminate $\omega(r)\,dr$. This expression integrates to

$$\overline{r^2} = \frac{3}{2b^2}$$

Since b^2 is $3/2\overline{r^2}$ [equation (5-29)] we get $\overline{r^2} = \overline{r^2}$.

5-4. This is just an extension of problem 5-1. Using the answers given there,

$$\overline{r^2} = l^2[1 + \cos \theta + \cos^2 \theta + \cos^3 \theta] +$$
$$l^2[\cos \theta + 1 + \cos \theta + \cos^2 \theta] +$$
$$l^2[\cos^2 \theta + \cos \theta + 1 + \cos \theta] +$$
$$l^2[\cos^3 \theta + \cos^2 \theta + \cos \theta + 1]$$

5-5. Equation (5-43) states

$$n_e = \frac{R^2}{\overline{r^2}} \quad \text{and} \quad l_e = \frac{\overline{r^2}}{R}$$

If the length of each monomer is 4.6 Å, R must be

$$R = 4.6 \text{ Å} n$$

where n is the number of monomer units. Thus,

$$n_e = \frac{(4.6n)^2}{16.2n} = 1.3n$$

and

$$l_e = \frac{16.2n}{4.6n} = 3.52 \text{ Å}$$

5-6. Equation (5-16) reads

$$\omega(m,n) = (\tfrac{1}{2})^n \frac{n!}{\left(\dfrac{n+m}{2}\right)! \left(\dfrac{n-m}{2}\right)!} \tag{5-16}$$

and if $n \gg m$ we can then use the form of Stirling's approximation given in the problem to get

$$\ln \omega(m,n) = (n + \tfrac{1}{2}) \ln n - \tfrac{1}{2}(n + m + 1) \ln\left[\left(\frac{n}{2}\right)\left(1 + \frac{m}{n}\right)\right]$$

$$- \tfrac{1}{2}(n - m + 1) \ln\left[\left(\frac{n}{2}\right)\left(1 - \frac{m}{n}\right)\right]$$

$$- \tfrac{1}{2} \ln 2\pi - n \ln 2$$

However, since $n \gg m$, we can use the series expansion

$$\ln\left(1 \pm \frac{m}{n}\right) = \pm \frac{m}{n} - \frac{m^2}{2n^2}$$

where third and higher order terms of (m/n) are neglected. We now have:

$$\ln \omega(m,n) = (n + \tfrac{1}{2}) \ln n - \tfrac{1}{2} \ln 2\pi - n \ln 2 - \tfrac{1}{2}(n + m + 1)$$

$$\times \left(\ln n - \ln 2 + \frac{m}{n} - \frac{m^2}{2n^2}\right) - \tfrac{1}{2}(n - m + 1)$$

$$\times \left(\ln n - \ln 2 - \frac{m}{n} - \frac{m^2}{2n^2}\right)$$

Simplifying and taking the anti-logarithm of both sides, we obtain

$$\omega(m,n) = \left(\frac{2}{\pi n}\right)^{\!\frac{1}{2}} \exp\left(-\frac{m^2}{2n}\right)$$

In the simplification we have made use of the fact that $n \gg 1$, so that

$$\frac{m^2}{2n}\left(\frac{1}{n} - 1\right) \approx -\frac{m^2}{2n}$$

5-7. Equation (5-32) is restated in this case as

$$\overline{r^2}' = \frac{\displaystyle\int_{1/2b}^{\infty} r^4 e^{-b^2 r^2} \, dr}{\displaystyle\int_{1/2b}^{\infty} r^2 e^{-b^2 r^2} \, dr}$$

where $\overline{r^2}'$ indicates a particular average quantity. Our problem then becomes one of evaluating the definite integrals given. This equation may be rewritten as

$$\overline{r^2}' = \frac{\displaystyle\int_{0}^{\infty} r^4 e^{-b^2 r^2} \, dr - \int_{0}^{1/2b} r^4 e^{-b^2 r^2} \, dr}{\displaystyle\int_{0}^{\infty} r^2 e^{-b^2 r^2} \, dr - \int_{0}^{1/2b} r^2 e^{-b^2 r^2} \, dr}$$

Now consider the second term in both the numerator and the denominator of this expression. If r is between 0 and $1/2b$, the exponential is quite close to 1. Thus we may approximate this expression as

$$\overline{r^2}' = \frac{\int_0^\infty r^4 e^{-b^2 r^2}\, dr - \int_0^{1/2b} r^4\, dr}{\int_0^\infty r^2 e^{-b^2 r^2}\, dr - \int_0^{1/2b} r^2\, dr}$$

which is integrated to give

$$\overline{r^2}' = \frac{3}{2b^2}\left[\frac{\sqrt{\pi} - \frac{1}{60}}{\sqrt{\pi} - \frac{1}{6}}\right]$$

Since the expression in brackets is greater than 1.0, $\overline{r^2}' > \overline{r^2}$. This result is clearly expected since conformations with small $\overline{r^2}$ have been excluded in calculating $\overline{r^2}'$.

5-8. Once again equation (5-32) is the starting point.

$$\bar{r} = \frac{\int_0^\infty r^3 e^{-b^2 r^2}\, dr}{\int_0^\infty r^2 e^{-b^2 r^2}\, dr}$$

In this case the variable transformation

$$y = b^2 r^2$$

is helpful. This yields

$$\bar{r} = \frac{1}{b}\frac{\int_0^\infty y e^{-y}\, dy}{\int_0^\infty y^{1/2} e^{-y}\, dy}$$

Both of these integrals are complete gamma functions so that

$$\bar{r} = \frac{1}{b}\frac{\Gamma(2)}{\Gamma(\frac{3}{2})} = \frac{1}{b}\frac{1 \cdot \Gamma(1)}{\frac{1}{2}\Gamma(\frac{1}{2})}$$

or

$$\bar{r} = \frac{2}{b\sqrt{\pi}}$$

In comparing the text with the results of this calculation we see that

$$(\overline{r^2})^{1/2} \neq \bar{r}$$

In addition, one should recall that $\bar{r}$ for a Gaussian chain is zero. These results are not inconsistent since the fact that makes the vector average zero is its sign, i.e., just as many $\mathbf{r}$ are positive as are negative along any axis.

In this problem, we have calculated the average of the radius of a spherical shell, a scalar quantity which cannot take on negative values.

Chapter 6

6-1(a)

$$dH = T\,dS + V\,dP + f\,dL$$

Thus

$$\left(\frac{\partial H}{\partial L}\right)_{T,P} = T\left(\frac{\partial S}{\partial L}\right)_{TP} + f$$

The Maxwell relation

$$\left(\frac{\partial S}{\partial L}\right)_{T,P} = -\left(\frac{\partial f}{\partial T}\right)_{L,P}$$

gives

$$\left(\frac{\partial H}{\partial L}\right)_{T,P} = -T\left(\frac{\partial f}{\partial T}\right)_{L,P} + f$$

Now

$$\left(\frac{\partial f}{\partial T}\right)_{P,L} = \left(\frac{\partial f}{\partial T}\right)_{P,\lambda} + \left(\frac{\partial f}{\partial \lambda}\right)_{P,L}\left(\frac{\partial \lambda}{\partial T}\right)_{PL}$$

Evaluating the various coefficients using equation (6-60) gives

$$\left(\frac{\partial f}{\partial T}\right)_{P,L} = \frac{\dot{f}}{T} - \frac{\alpha f}{3}\left(\frac{\lambda^3 + 2}{\lambda^3 - 1}\right)$$

and

$$\left(\frac{\partial H}{\partial L}\right)_{T,P} = \frac{\alpha T f}{3}\left(\frac{\lambda^3 + 2}{\lambda^3 - 1}\right)$$

(b)   Analogous to part (a) starting with dU instead of dH.

(c)

$$\left(\frac{\partial H}{\partial L}\right)_{T,P} = \frac{T\alpha}{\beta}\left(\frac{\partial V}{\partial L}\right)_{T,P}$$

Since

$$\left(\frac{\partial U}{\partial L}\right)_{T,V} = 0$$

Substituting the result of part (a)

$$\left(\frac{\partial V}{\partial L}\right)_{P,T} = \frac{\beta f}{3}\left(\frac{\lambda^3 + 2}{\lambda^3 - 1}\right)$$

Note that since $(\partial H/\partial L)_{T,P} \neq 0$, an ideal rubber has a thermoelastic inversion point.

6-2. Equation (6-11) defines the stress temperature coefficient in terms of pertinent quantities.

$$\left(\frac{\partial f}{\partial T}\right)_{P,L} = \frac{f - \left(\frac{\partial H}{\partial L}\right)_{T,P}}{T}$$

In problem 6-1a,

$$\left(\frac{\partial H}{\partial L}\right)_{T,P} = \frac{\alpha T f}{3}\left(\frac{\lambda^3 + 2}{\lambda^3 - 1}\right)$$

So

$$\left(\frac{\partial f}{\partial T}\right)_{P,L} = \frac{f\left[1 - \frac{\alpha T}{3}\left(\frac{\lambda^3 + 2}{\lambda^3 - 1}\right)\right]}{T}$$

6-3. Equation (6-42) in terms of this problem states

$$\Delta A = \tfrac{1}{3}NkT\overline{b^2 r^2}\left(\lambda^2 + \frac{1}{\lambda^2} - 2\right)$$

Incorporation of (6-44) gives

$$\Delta A = \tfrac{1}{2}NkT\frac{r_0^2}{r_f^2}\left(\lambda^2 + \frac{1}{\lambda^2} - 2\right)$$

Since pure shear is a constant volume process ($\lambda_1\lambda_2\lambda_3 = 1$)

$$\alpha^* = \lambda$$

and equation (6-52b) becomes

$$f = \frac{1}{L}\left(\frac{\partial A}{\partial \lambda}\right)_{T,V}$$

Operating with this on the second expression gives

$$f = \frac{NkT}{L}\frac{\overline{r_0^2}}{r_f^2}\left(\lambda - \frac{1}{\lambda^3}\right)$$

or

$$\sigma = N_0 kT \frac{\overline{r_0^2}}{r_f^2}\left(\lambda - \frac{1}{\lambda^3}\right)$$

6-4. According to equation (6-74)

$$I_1 = \lambda^2 + 1 + \frac{1}{\lambda^2}$$

and

$$I_2 = \lambda^2 + 1 + \frac{1}{\lambda^2}$$

Thus equation (6-80) states

$$\overline{W} = C_{100}\left(\lambda^2 + \frac{1}{\lambda^2} - 2\right) + C_{010}\left(\lambda^2 + \frac{1}{\lambda^2} - 2\right)$$

and using equation (6-79) gives

$$\sigma = 2[C_{100} + C_{010}]\left[\lambda - \frac{1}{\lambda^3}\right]$$

6-6. Substitution into equation (6-84) gives

$$G_0 = \frac{(0.95\ \text{gm/cc})(8.3 \times 10^7\ \text{erg/mole}^\circ\text{K})(300^\circ\text{K})}{5000\ \text{gm/mole}}\left(1 - \frac{10^4}{10^5}\right)$$

$$G_0 = 4.3 \times 10^6\ \text{dyne/cm}^2$$

Here we have assumed that the front factor $\overline{r_0^2}/r_f^2$ is equal to unity.

6-7. Using equation (6-94)

$$E_f = [1 + 2.5(0.3) + 14.1(0.3)^2] \, 5 \times 10^7 \text{ dynes/cm}^2$$

$$E_f = 1.5 \times 10^8 \text{ dynes/cm}^2$$

If half of the filler particles become ineffective,

$$V_f = 0.15$$

and

$$E_f = 8.3 \times 10^7 \text{ dynes/cm}^2$$

6-8. Starting with equation (6-66)

$$f = G_0 A_0 \left(\lambda - \frac{V}{V_0 \lambda^2} \right)$$

one can write

$$f = G_0 A_0 \left(\frac{L}{L_0} - \frac{V}{V_0} \frac{L_0^2}{L^2} \right)$$

Differentiation yields

$$\left(\frac{\partial f}{\partial T} \right)_{V,L} = \left(\frac{\partial G_0}{\partial T} \right)_{V,L} A_0 \left(\frac{L}{L_0} - \frac{V}{V_0} \frac{L_0^2}{L^2} \right)$$

$$+ \left(\frac{\partial A_0}{\partial T} \right)_{V,L} G_0 \left(\frac{L}{L_0} - \frac{V}{V_0} \frac{L_0^2}{L^2} \right)$$

$$+ G_0 A_0 \left[- \frac{L}{L_0^2} \left(\frac{\partial L_0}{\partial T} \right)_{V,L} + \frac{V}{V_0^2} \frac{L_0^2}{L^2} \left(\frac{\partial V_0}{\partial T} \right)_{V,L} \right.$$

$$\left. - 2 \frac{V}{V_0} \frac{L_0}{L^2} \left(\frac{\partial L_0}{\partial T} \right)_{V,L} \right]$$

Grouping terms and using the approximation

$$\frac{1}{V_0} \left(\frac{\partial V_0}{\partial T} \right)_{V,L} = \frac{3}{2} \frac{1}{A_0} \left(\frac{\partial A_0}{\partial T} \right)_{V,L} = 3 \frac{1}{L_0} \left(\frac{\partial L_0}{\partial T} \right)_{V,L}$$

leads directly to equation (6-67).

Chapter 7

7-1. Figure 7-3b is a plot of the relaxation behavior of a Maxwell body on a log-log scale. Thus we want to calculate

$$\frac{d \ln E(t)}{d \ln t}$$

where

$$E(t) = Ee^{-t/\tau}$$

However

$$\ln E(t) = \ln E - t/\tau$$

and since

$$\frac{d \ln E(t)}{d \ln t} = \frac{td \ln E(t)}{dt}$$

$$\frac{d \ln E(t)}{d \ln t} = -\frac{t}{\tau}$$

which approaches $-\infty$ for $t \gg \tau$.

7-2 and 7-3. As an example we will work with $E'(\omega)$ for a Maxwell body

(a)   Equation (2-56) gives

$$E'(\omega) = \omega \int_0^\infty \sin \omega s E(s) \, ds$$

$$E'(\omega) = \omega \int_0^\infty \sin \omega s E e^{-s/\tau} \, ds$$

which integrates to

$$E'(\omega) = \frac{E\omega^2\tau^2}{1 + \omega^2\tau^2}$$

(b)   Starting with equation (7-5) we find

$$E\varepsilon_0 \int_{t_1}^{t_2} e^{t/\tau} \cos \omega t \, dt = \int_{\sigma(t_1)}^{\sigma(t_2)} d(\sigma e^{t/\tau})$$

assuming that

$$\varepsilon(t) = \varepsilon_0 \sin \omega t$$

This expression integrates to

$$\sigma(t_2) - \sigma(t_1)e^{\frac{t_1-t_2}{\tau}} = \frac{E\omega\varepsilon_0}{\frac{1}{\tau^2} + \omega^2}\left[\frac{1}{\tau} \cos \omega t_2 + \omega \sin \omega t_2\right.$$

$$\left. - e^{\frac{t_1-t_2}{\tau}}\left(\frac{1}{\tau} \cos \omega t_1 + \omega \sin \omega t_1\right)\right]$$

Since our result must be independent of the starting time t_1, and the present time t_2 ($E'(\omega)$ is not a function of t) let $t_2 \to \infty$. Then

$$\sigma(t_2) = \frac{E\omega\varepsilon_0}{\frac{1}{\tau^2} + \omega^2} \left[\frac{1}{\tau} \cos \omega t_2 + \omega \sin \omega t_2\right]$$

$E'(\omega)$ however measures only the in phase stress so

$$E'(\omega) = \frac{\sigma'(t_2)}{\varepsilon(t_2)} = \frac{E\omega\varepsilon_0}{\frac{1}{\tau^2} + \omega^2} \left[\frac{\omega \sin \omega t_2}{\varepsilon_0 \sin \omega t_2}\right]$$

or

$$E'(\omega) = \frac{E\omega^2\tau^2}{1 + \omega^2\tau^2}$$

7-4. Here we can start with equation (1) of example 7-2 with added subscripts.

$$\sigma_i(t_2) = \frac{E_i\omega\varepsilon_0}{\frac{1}{\tau_i^2} + \omega^2} \left[\frac{1}{\tau_i} \cos \omega t_2 + \omega \sin \omega t_2\right]$$

This expression can be separated into an in phase σ_i' part and an out of phase σ_i'' part as follows:

$$\sigma_i'(t_2) = \frac{E_i\omega\varepsilon_0}{\frac{1}{\tau_i^2} + \omega^2} [\omega \sin \omega t_2]$$

$$\sigma_i''(t_2) = \frac{E_i\omega\varepsilon_0}{\frac{1}{\tau_i^2} + \omega^2} \left[\frac{1}{\tau_i} \cos \omega t_2\right]$$

Now the total in phase stress considering all elements is just

$$\sigma'(t_2) = \sum_i \frac{E_i\omega^2\varepsilon_0}{\frac{1}{\tau_i^2} + \omega^2} (\sin \omega t_2)$$

Here the quantity in brackets merely expresses the phase of the response (in phase) and we may write immediately

$$E'(\omega) = \sum_i \frac{E_i\omega^2}{\dfrac{1}{\tau_i^2} + \omega^2} = \sum_i \frac{E_i\omega^2\tau_i^2}{1 + \omega^2\tau_i^2}$$

Similarly for the out of phase response,

$$\sigma''(t_2) = \sum_i \frac{E_i\omega\varepsilon_0}{\dfrac{1}{\tau_i^2} + \omega^2}\left[\frac{1}{\tau_i}\cos\omega t_2\right]$$

Here $\cos\omega t_2$ just stipulates out of phase response and

$$E''(\omega) = \sum_i \frac{E_i\omega\tau_i}{1 + \omega^2\tau_i^2}$$

Then according to equation (2-22)

$$E^*(\omega) = \sum_i \frac{E_i\omega^2\tau_i^2}{1 + \omega^2\tau_i^2} + i\sum_i \frac{E_i\omega\tau_i}{1 + \omega^2\tau_i^2}$$

Clearly such sums are difficult to invert analytically.

7-5. Here we can write

$$\frac{d\varepsilon_1}{dt} = \frac{1}{E_1}\frac{d\sigma_1}{dt} + \frac{\sigma_1}{\eta_1}$$

$$\frac{d\varepsilon_i}{dt} = \frac{1}{E_i}\frac{d\sigma_i}{dt} + \frac{\sigma_i}{\eta_i}$$

$$\frac{d\varepsilon_z}{dt} = \frac{1}{E_z}\frac{d\sigma_z}{dt} + \frac{\sigma_z}{\eta_z}$$

These expressions simplify considerably since

$$\sigma_1 = \sigma_2 \cdots = \sigma_i = \cdots \sigma_z$$

and

$$\sum_{i=1}^{z}\frac{d\varepsilon_i}{dt} = \frac{d\varepsilon}{dt}$$

Thus

$$\frac{d\varepsilon}{dt} = \frac{d\sigma}{dt}\sum_{i=1}^{z}\frac{1}{E_i} + \sigma\sum_{i=1}^{z}\frac{1}{\eta_i}$$

Defining

$$\frac{1}{\mathscr{E}} = \sum_{i=1}^{z} \frac{1}{E_i}$$

and

$$\frac{1}{H} = \sum_{i=1}^{z} \frac{1}{\eta_i}$$

this equation becomes

$$\frac{d\varepsilon}{dt} = \frac{1}{\mathscr{E}} \frac{d\sigma}{dt} + \frac{\sigma}{H}$$

which is solvable in the usual way.

7-6(a). Equation (7-70) can be rearranged to give

$$\rho a^2 z^2 = \frac{36}{N} (\eta)$$

Since all of the terms on the left hand side are parameters relating to the model itself and not to the specific experimental conditions, this result may be substituted into the expression for τ_p in the problem to give the desired result.

(b) Table 7-2 gives an expression for the tensile creep compliance of a Voigt-Kelvin Model as

$$D(t) = \sum_{p=1}^{z} D_p(1 - e^{-t/\tau_p})$$

Comparison of this expression with that in the problem yields the result that

$$D_p = \frac{8}{3NkT\pi^2} \frac{1}{p^2}$$

7-7. First let us investigate the values of $E''(\omega)$ and ω at the maximum of a log $E''(\omega)$ versus log ω curve for a single Maxwell element.

$$\log E''(\omega) = \log E + \log \omega + \log \tau - \log(1 + \omega^2\tau^2)$$

$$\frac{d \log E''(\omega)}{d \log \omega} = 1 - \frac{2\omega^2\tau^2}{1 + \omega^2\tau^2}$$

At the maximum this slope is zero and ω is found to equal $1/\tau$. Substitution of this result into the defining equation for $E''(\omega)$ shows that

$$E''(\omega) = E/2$$

at the maximum. Thus

$$\tau_1 = 10^{-2} \quad \text{and} \quad E_1 = 4 \times 10^6$$
$$\tau_2 = 10^{-9} \quad \text{and} \quad E_2 = 2 \times 10^{10}$$

7-8.

$$\eta = \int_0^\infty G(t)\,dt \tag{3-7}$$

For a Maxwell-Weichert Model

$$\eta = \int_0^\infty \sum_{i=1}^z G_i e^{-t/\tau_i}\,dt$$

or

$$\eta = \sum_{i=1}^z G_i \int_0^\infty e^{-t/\tau_i}\,dt$$

$$= \sum_{i=1}^z G_i \tau_i$$

7-9.

$$E(t) = \sum_{p=1}^z E_p e^{-t/\tau_p}$$

$$\approx E_p \int_{p=1}^z e^{-tp/\tau_{\max}}\,dp$$

$$\approx E_p \frac{\tau_{\max}}{t}[e^{-t/\tau_{\max}} - e^{-t/\tau_{\min}}]$$

If $\tau_{\min} < t < \tau_{\max}$, the quantity in brackets is close to 1 and relatively constant. Thus,

$$\frac{d\ln E(t)}{d\ln t} = -1$$

Chapter 8

8-2. Binomial expansion yields

$$\left(1 - \frac{q}{N(0)x}\right)^x = 1 - \frac{xq}{N(0)x} + \frac{(x)(x-1)}{2!}\left(\frac{q}{N(0)x}\right)^2$$
$$- \frac{(x)(x-1)(x-2)}{3!}\left(\frac{q}{N(0)x}\right)^3 \cdots$$
$$\approx 1 - \frac{q}{N(0)} + \frac{1}{2!}\left(\frac{q}{N(0)}\right)^2 - \frac{1}{3!}\left(\frac{q}{N(0)}\right)^3$$

which is just the expansion of the exponential $e^{-q/N(0)}$. Thus

$$\left(1 - \frac{q}{N(0)x}\right)^x \approx e^{-q/N(0)}$$

8-3. Start with equation (8-10) which states

$$q = -N(0) \ln \frac{\sigma(t)}{\sigma(0)}$$

The ln function can be expanded according to the series

$$\ln x = \frac{x-1}{x} + \frac{1}{2}\left(\frac{x-1}{x}\right)^2 + \frac{1}{3}\left(\frac{x-1}{x}\right)^3 \cdots$$

But if x is close to 1

$$\ln x \approx \frac{x-1}{x}$$

Thus

$$q \approx -N(0)\left(\frac{\dfrac{\sigma(t)}{\sigma(0)} - 1}{\dfrac{\sigma(t)}{\sigma(0)}}\right)$$

$$q = -N(0)\left(\frac{\sigma(0)}{\sigma(t)} - 1\right) \tag{8-20}$$

8-4. This can be done by simply writing several terms of the binomial expansion

$$\left(1 - \frac{1}{N(0)}\right)^q = 1 - \frac{q}{N(0)}$$
$$+ \frac{q(q-1)}{2!}\left(\frac{1}{N(0)}\right)^2 - \frac{q(q-1)(q-2)}{3!}\left(\frac{1}{N(0)}\right)^3 \cdots$$

8-5. This summation may be written explicitly as

$$N(0)p[1 + 2(1 - p) + 3(1 - p)^2 + 4(1 - p)^3 \cdots]$$

Consider the series in brackets and replace $1 - p$ by y. Thus we desire the sum

$$1 + 2y + 3y^2 + 4y^3 \cdots$$

The MacLaurin expansion of $1/(1 - y)^2$ generates just this sum. Working back through the variable transform,

$$N(0)p\frac{1}{p^2} = \frac{N(0)}{p}$$

the result quoted in equation (8-28).

8-6. From equation (8-37),

$$\frac{N_u}{N_x}\left[\left(\frac{L_s}{L_u}\right)^2 - \frac{L_u}{L_s}\right] = \frac{L_x}{L_s} - \left(\frac{L_s}{L_x}\right)^2$$

the following rearrangements can be made:

$$\frac{N_u}{N_x}\left[\frac{L_s^3 - L_u^3}{L_u^2 L_s}\right] = \frac{L_x^3 - L_s^3}{L_x^2 L_s}$$

$$\frac{N_u L_x^2}{N_x L_u^2} = \frac{(L_x^3/L_u^3) - (L_s^3/L_u^3)}{(L_s^3/L_u^3) - 1} = \frac{(L_x^3/L_u^3) - 1}{(L_s^3/L_u^3) - 1} - 1$$

$$\frac{L_s^3}{L_u^3} - 1 = \frac{(L_x^3/L_u^3) - 1}{(N_u L_x^2/N_x L_u^2) + 1}$$

$$\frac{L_s}{L_u} = \left[\frac{(L_x^3/L_u^3) - 1}{(N_u L_x^2/N_x L_u^2) + 1} + 1\right]^{1/3}$$

The permanent set is defined by equation (8-33):

$$p.s. = \frac{L_s - L_u}{L_x - L_u} \times 100$$

$$= \frac{(L_s/L_u) - 1}{(L_x/L_u) - 1} \times 100$$

$$= \frac{100}{(L_x/L_u) - 1} \left\{ \left[\frac{(L_x^3/L_u^3) - 1}{(N_u L_x^2/N_x L_u^2) + 1} + 1 \right]^{\frac{1}{3}} - 1 \right\}$$

$$= C_3 \left\{ \left[\frac{C_1}{(N_u/N_x)C_2 + 1} + 1 \right]^{\frac{1}{3}} - 1 \right\}$$

where C_1, C_2, and C_3 are defined as in the text.

8-7. In general a rate constant can be written as

$$k = Ae^{-H_a/RT}$$

where all of the temperature dependence is embodied in the exponential term. If one performs the operation

$$RT^2 \frac{d \ln k}{dT}$$

the activation energy is extracted. Applying this procedure to equation (8-50) yields (8-51).

List of Major Symbols

◆◆◆◆◆◆◆◆◆◆◆◆◆◆◆◆◆◆◆◆◆◆◆◆◆◆◆◆◆◆◆◆◆◆◆◆◆

Certain symbols in this book were used to designate different meanings. This redundancy is preferred in order to preserve conformity with those commonly accepted in the literature. Those symbols the usage of which are restricted to certain chapters are so indicated. Some minor symbols are defined in the text where applicable, and are not included in this list.

A	Helmholtz free energy
A_0	cross-sectional area
$[A]$	Zimm matrix
a	entanglement parameter
a^2	mean square end-to-end distance of submolecule
a_T	shift factor
B	$3kT/a^2\rho$ (Chapter 7)
b^2	$3nl^2/2$
C_P	constant pressure heat capacity
C_V	constant volume heat capacity
C_1, C_2	WLF parameters
C_1, C_2	Mooney-Rivlin parameters (Chapter 6)
C_1, C_2, C_3	permanent set parameters (Chapter 8)
c	concentration
c	Fox-Flory constant (Chapter 4)
D	tensile compliance
$D(t)$	tensile creep compliance
d	density
E, E_0	tensile modulus
$E(t)$	tensile relaxation modulus
E^*	complex tensile modulus
E'	dynamic storage tensile modulus
E''	dynamic loss tensile modulus

E_e	equilibrium tensile modulus
E_f	tensile modulus of filled rubber
E_1	glassy tensile modulus
E_2	rubbery tensile modulus
e_s	efficiency of scission
F	Gibbs free energy
f	force
f	fractional free volume (Chapter 3)
f_e	energetic stress
f_g	fractional free volume at glass transition
f_s	entropic stress
G, G_0	shear modulus
$G(t)$	shear relaxation modulus
G^*	complex shear modulus
G'	dynamic storage shear modulus
G''	dynamic loss shear modulus
G_1	glassy shear modulus
G_2	rubbery shear modulus
H	enthalpy
H_a	activation energy
$H(\tau)$	distribution of relaxation times
I_1, I_2, I_3	strain invariants
$[I]$	identity matrix
I_0	initiator concentration
i	$\sqrt{-1}$
J	shear compliance
$J(t)$	shear creep compliance
J^*	complex shear compliance
J'	dynamic storage shear compliance
J''	dynamic loss shear compliance
J_r	recoverable compliance
K_A, K_B	Gordon-Taylor parameters
k	Boltzmann constant
k	K_A/K_B (Chapter 4)
k_d	rate constant for initiator decomposition
k_i	rate constant for initiation
k_p	rate constant for propagation
k_t	rate constant for termination

L	length of strained sample
L	Laplace transform (Chapter 2)
L_0	length of unstrained sample at V_0
L'	length of unstrained sample at V
L_s	new unstrained length
L_u	original unstrained length
L_x	strained length
$L(\tau)$	distribution of retardation times
l	length of a chain link
$\vec{l}$	vectorial length of a chain link
M	modulus (Chapter 1)
M	concentration of molecular segments (Chapter 8)
M	molecular weight
M_c	molecular weight of network chain (Chapter 6)
M_c	critical entanglement molecular weight (Chapter 7)
$\bar{M}_n$	number average molecular weight
m	difference between positive and negative steps (Chapter 5)
m	number of polysulfide links (Chapter 8)
N	number of network chains
N_{Av}	Avogadro's number
N_0	number of network chains per unit volume
N_u	original chain concentration
N_x	chain concentration corresponding to permanent set length
$\mathfrak{N}$	fixed number of conformations of chain
n	number of links in a polymer chain
n_+	number of positive steps
n_-	number of negative steps
P	pressure
p	size parameter
$[Q]$	transformation matrix
q	number of cuts per unit volume (Chapter 8)
q	elements of $[q]$ matrix (Chapter 7)
$\dot{q}$	time derivative of q
$[q]$	normal coordinate matrix
R	extend chain length (Chapter 5)
R	ideal gas constant

$\overline{r^2}$	mean square end-to-end distance
$\vec{r}$	vectorial length of chain ends
$\overline{r_f^2}$	mean square end-to-end distance of a free chain
$\overline{r_0^2}$	mean square end-to-end distance of a real chain
S	entropy
s	negative slope in the transition region of modulus-temperature curve
T	temperature
T_g	glass transition temperature
T_i	inflection temperature
T_m	melting temperature
T_{ref}	reference temperature
T_2	second order transition temperature
t	time
t_{ref}	reference time
U	internal energy
u_0	hole formation energy
V	volume
V_d	volume fraction of diluent
V_f	free volume
V_f	volume fraction of filler (Chapter 6)
V_0	initial volume
V_p	volume fraction of polymer
V_r	volume fraction of rubber
W	work
$\overline{W}$	strain energy
W_A, W_B	weight fraction of comonomers
X	displacement
$\dot{X}$	time derivative of displacement
X_A, X_B	mole fractions of comonomers
x	number of segments per chain
Z	number of primary valences of chain atom
z	number of mechanical elements or submolecules
α	thermal expansion coefficient
α^*	extension ratio at constant volume
α_f	thermal expansion coefficient of free volume
α_g	thermal expansion coefficient of glass
α_r	thermal expansion coefficient of rubber

β	isothermal compressibility
β_f	isothermal compressibility of free volume
Γ	gamma function
γ	shear strain
$\vec{\gamma}$	strain vector
$\vec{\gamma}'$	in-phase component of strain vector
$\vec{\gamma}''$	out-of-phase component of strain vector
Δ	logarithmic decrement
δ	loss angle
ε	tensile strain
$\varepsilon, \varepsilon_A, \varepsilon_B$	flex energies (Chapter 4)
η	viscosity
θ	$\pi - \psi$
Λ	extension ratio including strain amplification factor
$[\Lambda]$	diagonalized Zimm matrix
λ	extension ratio at constant pressure
$\lambda_1, \lambda_2, \lambda_3$	extension ratios in principal directions
λ_c	extension ratio of dry rubber
λ_p	elements of $[\lambda]$ matrix
λ_s	extension ratio of swollen rubber
$\lambda_1, \lambda_2, \lambda_3$	principal extension ratios
μ	Poisson ratio
π	180° angle or 3.1416
ρ	density (Chapter 3)
ρ	segmental friction factor (Chapter 7)
ρ_0	segmental friction factor at short times
$\vec{\sigma}$	stress vector
$\vec{\sigma}'$	in-phase component of stress vector
$\vec{\sigma}''$	out-of-phase component of stress vector
σ_s	shear stress
σ_t	tensile stress
σ'	true stress
$\sigma_1, \sigma_2, \sigma_3$	principal stresses
τ	relaxation or retardation time
τ_c	critical relaxation time
τ_{min}	minimum relaxation time
τ_p	relaxation time of Rouse model
τ_1	maximum relaxation time

$[\phi]$	transformation matrix
ψ	bond angle
Ω	number of conformations
ω	angular frequency (Chapters 2, 7)
ω	probability (Chapters 5, 6)

Author Index
◆◆◆◆◆◆◆◆◆◆◆◆◆◆◆◆◆◆◆◆◆◆◆

245

Subject Index

◆◆◆◆◆◆◆◆◆◆◆◆◆◆◆◆◆◆

247